RULE OF
THE
ROBOTS

ALSO BY MARTIN FORD

The Lights in the Tunnel:
Automation, Accelerating Technology
and the Economy of the Future
(Acculant Publishing, 2009)

Rise of the Robots:
Technology and the Threat of a Jobless Future
(Basic Books, 2015)

Architects of Intelligence:
The Truth about AI from the People Building It
(Packt Publishing, 2018)

RULE OF THE ROBOTS

HOW ARTIFICIAL INTELLIGENCE WILL TRANSFORM EVERYTHING

MARTIN FORD

BASIC BOOKS

New York

Basic Books
Hachette Book Group
1290 Avenue of the Americas, New York, NY 10104
www.basicbooks.com

Printed in the United States of America

Published by Basic Books, an imprint of Perseus Books, LLC, a subsidiary of Hachette Book Group, Inc. The Basic Books name and logo is a trademark of the Hachette Book Group.

The Hachette Speakers Bureau provides a wide range of authors for peaking events. To find out more, go to www.hachettespeakersbureau.com or call (866) 376-6591.

The publisher is not responsible for websites (or their content) that are not owned by the publisher.

Print book interior design by Jeff Williams.

Library of Congress Cataloging-in-Publication Data

Names: Ford, Martin (Martin R.) author.
Title: Rule of the robots: how artificial intelligence will transform everything / Martin Ford.
Description: New York: Basic Books, [2021] | Includes bibliographical references and index. |
Identifiers: LCCN 2021016340 | ISBN 9781541674738 (hardcover) | ISBN 9781541674721 (ebook)
Subjects: LCSH: Artificial intelligence—Social aspects. | Robotics—Social aspects.
Classification: LCC Q334.7 .F67 2021 | DDC 006.301—dc23

LC record available at https://lccn.loc.gov/2021016340

ISBNs: 978-1-5416-7473-8 (hardcover); 978-1-5416-7472-1 (ebook)

LSC-C

Printing 1, 2021

For my mother, Sheila

CONTENTS

CHAPTER 1

THE EMERGING DISRUPTION

ON NOVEMBER 30, 2020, DEEPMIND, A LONDON-BASED ARTIFICIAL intelligence company owned by Google parent Alphabet, announced a stunning, and likely historic, breakthrough in computational biology, an innovation with the potential to genuinely transform science and medicine. The company had succeeded in using deep neural networks to predict how a protein molecule will fold into its final shape based on the genetic code from which the molecule is constructed in cells. It was a milestone that culminated a fifty-year scientific quest and marked the advent of a new technology that was poised to usher in an unprecedented understanding of the very fabric of life—as well as a new age of medical and pharmaceutical innovation.[1]

Protein molecules are long chains in which each link consists of one of twenty different amino acids. The genes encoded in DNA lay out the precise sequence, or essentially the recipe, of the amino acids that make up the protein molecule. This genetic recipe, however, does not specify the shape of the molecule, which is critical to its function. Instead, the shape results from the way the molecule automatically folds into a highly complex three-dimensional structure within milliseconds of its fabrication in the cell.[2]

Predicting the exact configuration into which a protein molecule will fold is one of the most daunting challenges in science. The number of possible shapes is virtually infinite. Scientists have devoted entire careers to the problem, but have collectively achieved only modest success. DeepMind's system uses AI techniques that the company pioneered in the AlphaGo and Alpha-Zero systems that had famously triumphed over the world's best human competitors at board games like Go and chess. But the era of AI being primarily associated with adeptness at games is clearly drawing to a close. AlphaFold's ability to predict the shape of protein molecules with an accuracy that rivals expensive and time-consuming laboratory measurement using techniques like X-ray crystallography offers irrefutable evidence that research at the very frontier of artificial intelligence has produced a practical and indispensable scientific tool with the potential to transform the world.

Arriving at a moment when nearly everyone on earth had likely encountered illustrations featuring the most notorious example of how a protein molecule's three-dimensional shape defines its function—the coronavirus spike protein, a kind of molecular docking mechanism that allows the virus to attach to and infect its host—the breakthrough offered hope that we will be far better prepared for the next pandemic. One important use of the system might be to rapidly screen existing medications to find the ones likely to be most effective against a newly emergent virus, putting powerful treatments in the hands of doctors in the earliest stages of an outbreak. Beyond this, DeepMind's technology is poised to lead to a variety of advances, including the design of entirely new drugs and a better understanding of the ways in which proteins can misfold—something that has been associated with illnesses like diabetes as well as Alzheimer's and Parkinson's diseases. The technology might someday be employed in a range of applications outside medicine, for example, to help engineer microbes that can secrete proteins capable of degrading waste such as plastic or oil.[3] In other words, it is an innovation with the potential to accelerate progress in virtually every sphere of biochemical science and medicine.

Over roughly the past decade, the field of artificial intelligence has taken a revolutionary leap forward and is beginning to deliver an ever-increasing number of practical applications that are already transforming the world around us. The primary accelerant of this progress has been "deep learning"—a machine learning technique based on the use of multilayered artificial neural networks of the kind employed by DeepMind. The basic principles of deep neural networks have been understood for decades, but recent dramatic advances have been enabled by the confluence of two relentless trends in information technology: First, the arrival of vastly more powerful computers has, for the first time, allowed neural networks to transition into truly capable tools. And, second, the enormous troves of data now being generated and collected across the information economy provide a resource crucial to training these networks to perform useful tasks. Indeed, the availability of data at a scale that would have once been unimaginable is arguably the single most important factor underlying the startling progress we have seen. Deep neural networks hoover up and leverage data much in the way that a massive blue whale feeds on tiny krill, scooping up vast numbers of individually insignificant organisms and then using their collective energy to animate a creature of magnificent size and power.

As artificial intelligence is successfully applied to more and more areas, it is becoming clear that it is evolving into a uniquely consequential technology. In some specific areas of medicine, for example, diagnostic AI applications are already beginning to match or even exceed the performance of the best doctors. The true power of such an innovation does not lie just in its ability to potentially outperform a single world-class physician, but rather in the ease with which the intelligence encapsulated in the technology can be scaled. Someday soon, elite diagnostic expertise will be affordably broadcast across the globe, making it available even in regions where people barely have access to any doctor or nurse—let alone to one of the world's best medical specialists.

Now imagine taking a single, extremely specific innovation—an AI-based diagnostic tool or perhaps Deep-Mind's breakthrough in protein folding—and multiplying it by a virtually limitless number of possibilities in other areas from medicine to science, industry, transportation, energy, government and every other sphere of human activity. What you end up with is a new, and uniquely powerful, utility. In essence, an "electricity of intelligence." A flexible resource that can— perhaps someday with almost a flick of a switch—apply cognitive capability to virtually any problem we face. Ultimately, this new utility will deliver the ability not just to analyze and make decisions but to solve complex problems and even exhibit creativity.

The purpose of this book will be to explore the future implications of artificial intelligence by viewing it not as a specific innovation, but rather as a uniquely scalable and potentially disruptive technology—a powerful new utility poised to deliver a transformation that will someday rival the impact of electricity. The arguments and explanations I will put forth here draw heavily on three of my own professional experiences.

First, since the publication of my book *Rise of the Robots: Technology and the Threat of a Jobless Future* in 2015, I have been invited to speak about the impact of artificial intelligence and robotics at dozens of technology conferences, regional summits and corporate and academic events. I've traveled to more than thirty countries and have had an opportunity to visit research labs, to see demonstrations of leading-edge technology and to discuss and debate the implications of the unfolding AI revolution with technical experts, economists, business executives, investors and politicians, as well as average people who are seeing—and beginning to worry about—the changes happening around them.

Second, in 2017 I began working with a team at the French bank Société Générale to create a proprietary stock market index that would offer investors a way to benefit directly from the artificial intelligence and robotics revolution. In my role as the consulting thematic expert, I helped formulate a strategy informed by the view that AI is becoming a powerful new utility and that it will therefore generate value and transform businesses in a wide range of industries. The result was Société Générale's "Rise of the Robots" index and subsequently the Lyxor Robotics and AI ETF[4] (exchange traded fund), which is based on the index.

Finally, throughout 2018, I had an opportunity to sit down and have wide-ranging discussions with twenty-three of the world's foremost artificial intelligence research scientists and entrepreneurs. These men and women are truly the "Einsteins" of the field, and indeed, four of the people I spoke with have won the Turing award, computer science's equivalent of the Nobel prize. These conversations, which delved into the future of artificial intelligence as well as the risks and opportunities that progress will bring, are recorded in my 2018 book *Architects of Intelligence: The Truth about AI from the People Building It*. I have drawn extensively from this unique opportunity to get inside some of the absolute brightest minds working in the field of AI, and their insights and predictions directly inform much of the material in this book.

Viewing artificial intelligence as the new electricity offers a useful model for thinking about how the technology will evolve and ultimately touch nearly every sphere of the economy, society and culture. However, there is one important caveat. Electricity is universally viewed as an unambiguously positive force. Setting aside the most dedicated hermit, it would probably be hard to find anyone living in a developed country who has reason to regret electrification. AI is different: it has a dark side, and it comes coupled with genuine risks both to individuals and to society as a whole.

As artificial intelligence continues to advance, it has the potential to upend both the job market and the overall economy to a degree that is likely unprecedented. Virtually any job that is fundamentally routine and predictable in nature—or in other words, nearly any role where a worker faces similar challenges again and again—has the potential to be automated in full or in part. Studies have found that as much as half of the American workforce is engaged in such predictable activities, and that tens of millions of jobs could eventually evaporate in the United States alone.[5] And the impact will not be limited to lower-wage, unskilled workers. Many people in white collar and professional roles likewise perform relatively routine tasks. Predictable intellectual work is at especially high risk of automation because it can be performed by software. Manual labor, in contrast, requires an expensive robot.

There continues to be a vibrant debate over the impact of automation on the future workforce. Will sufficient new, nonautomatable jobs be created to absorb the workers who lose more routine work? And, if so, will these workers have the necessary skills, capabilities and personality traits to successfully transition into these newly created roles? We probably should not assume that most former truck drivers or fast food workers can become robotics engineers—or, for that matter, personal care assistants for the elderly. My own view, as I argued in *Rise of the Robots*, is that a large fraction of our workforce is eventually at risk of being left behind as AI and robotics continue to advance. And, as we'll see, there are very good reasons to believe that the coronavirus pandemic and the associated economic downturn will accelerate the impact of artificial intelligence on the job market.

Even if we set aside the complete elimination of jobs through automation, technology is already affecting the job market in other ways that should concern us. Middle class jobs are at risk of being deskilled, so that a low-wage worker with little training, but who is augmented by technology, can step into a role that once would have commanded a higher wage. People are increasingly working under the control of algorithms that monitor or pace their work, in effect treating them like virtual robots. Many of the new opportunities being created are in the "gig" economy, where workers typically have unpredictable hours and incomes. All of this points to increasing inequality and potentially dehumanizing conditions for a growing fraction of our workforce.

Aside from the impact on jobs and the economy, there are a variety of other dangers that will accompany the continuing rise of artificial intelligence. One of the most immediate threats will be to our overall security. This includes AI-enabled cyberattacks on physical infrastructure and critical systems that increasingly will be interconnected and managed by algorithms, as well as threats to the democratic process and the social fabric. The Russian intervention in the 2016 presidential election offers a relatively tame preview of what might be coming. Artificial intelligence could eventually put "fake news" on steroids by enabling the creation of photographic, audio and video fabrications that are virtually indistinguishable from reality, while armies of truly advanced bots could someday invade social media, sow confusion and potentially mold public opinion with terrifying proficiency.

Throughout the world—but especially in China—surveillance systems employing facial recognition and other AI-based technologies are being used in ways that vastly enhance the power and reach of authoritarian governments and erode any expectation of personal privacy. In the United States, facial recognition systems have, in some cases, been shown to be biased on the basis of race or gender, as have algorithms used to screen resumes or even to advise judges acting within the criminal justice system.

Perhaps the most terrifying near-term threat is the development of fully autonomous weapons with the ability to kill without the necessity of a human giving specific authorization. Such weapons could conceivably be used en masse to target entire populations and would be extremely hard to defend against, especially if they fell into the hands of terrorists. This is a development that many people in the AI research community are passionate about preventing, and there is an initiative underway at the United Nations to ban such weapons.

Further in the future, we may encounter an even greater danger. Could artificial intelligence pose an existential threat to humanity? Could we someday build a "superintelligent" machine, something so far beyond us in its capability that it might, either intentionally or inadvertently, act in ways that cause us harm? This is a far more speculative fear that arises only if we someday succeed in building a genuinely intelligent machine. This remains the stuff of science fiction. Nonetheless, the quest to create true, human-level artificial intelligence is the Holy Grail of the field, and a number of very smart people take this concern very seriously. Prominent individuals like the late Stephen

Hawking and Elon Musk have issued warnings about the specter of out-of-control AI, with Musk in particular setting off a media frenzy by declaring that artificial intelligence research is "summoning the demon" and that "AI is more dangerous than nuclear weapons."[6]

Given all this, one might wonder why we should choose to open Pandora's box. The answer is that humanity cannot afford to leave artificial intelligence on the table. Because AI will amplify our intellect and creativity, it will drive innovation across virtually every field of human endeavor. We can anticipate new drugs and medical treatments, more efficient clean energy sources and a multitude of other important breakthroughs. AI will certainly destroy jobs, but it will also make the products and services produced by the economy more affordable and available. An analysis from the consulting firm PwC predicts that AI will add about $15.7 trillion to the global economy by the year 2030—and this is all the more critical as we look forward to recovery from the massive economic crisis unleashed by the coronavirus pandemic.[7] Perhaps most importantly, artificial intelligence will evolve into an indispensable tool that will be crucial in addressing the greatest challenges we face, including climate change and environmental degradation, the inevitable next pandemic, energy and fresh water scarcity, poverty and lack of access to education.

The path forward must be to fully embrace the potential of artificial intelligence, but to do so with open eyes. The risks will need to be addressed. Specific applications of AI will need to be regulated and, in some instances, banned. All this needs to begin happening now because the future is poised to arrive long before we are ready for it.

To claim that this book will offer a "roadmap" to the future of artificial intelligence would be to engage in hyperbole. No one knows how rapidly AI will advance, the specific ways in which it will be leveraged, the new companies and industries that will arise or the dangers that will loom largest. The future of artificial intelligence is likely to be as unpredictable as it is disruptive. There is no roadmap. We will have to think on our feet. My hope is that this book will offer a way to prepare for what is to come: a guide to thinking about the unfolding revolution, separating hype and sensationalism from reality, and identifying the best ways for both individuals and our society as a whole to thrive in the future we are creating.

CHAPTER 2

AI AS THE NEW ELECTRICITY

ELECTRICITY, A FORCE THAT WAS ONCE VALUED SOLELY AS A SOURCE of entertainment in crowd-pleasing tricks and experiments, has indisputably shaped and enabled modern civilization. In a world where guaranteed access to the electrical grid is often taken for granted, it is easy to forget just how long and arduous electricity's climb to dominance actually was. From Benjamin Franklin's famous kite experiment in 1752, a full 127 years passed before Thomas Edison finally perfected his incandescent light bulb in 1879. From that point, things moved faster. That same year in the United Kingdom, the Liverpool Electric Lighting Act laid the groundwork for the country's first electric street lighting, and just three years later, both the Pearl Street Power Plant in New York City and the Edison Electric Light Station in London began operating. Still, by 1925, only about half of homes in the United States had access to electric power. It took several more decades and Franklin Roosevelt's Rural Electrification Act before electricity evolved into the ubiquitous utility that we know today.

For those of us who live in the developed world, there is virtually nothing that is not somehow touched by, or indeed made possible by, access to electric power. Electricity is probably the best—and certainly the most durable—example of a general-purpose technology: in other words, an innovation that scales across and transforms every aspect of the economy and society. Other general-purpose technologies include steam power, which produced the Industrial Revolution, but is now relegated to a few applications like nuclear power plants. The internal combustion engine was certainly transformative, but it's now quite easy to imagine a future where gas and diesel engines are almost entirely displaced—likely by electric motors. In the absence of some dystopian catastrophe scenario, it's almost impossible to imagine a future without electricity.

It is, therefore, an extraordinarily bold claim to argue that artificial intelligence will evolve into a general-purpose technology of such scale and power that it can reasonably be compared to electricity. Nonetheless, there are good reasons to believe that this is the path we are on: AI, much like electricity, will eventually touch and transform virtually everything.

Artificial intelligence is already impacting every sector of the economy, including agriculture, manufacturing, healthcare, finance, retail and virtually all other industries. The technology is even beginning to invade areas that we consider the most human. Already, AI-enabled chatbots provide round-the-clock access to mental health counseling. New forms of graphic art and music are being generated with deep learning technology. None of this should really surprise us. After all, virtually everything of value that human beings have created is a direct product of our intelligence—of our ability to learn, to innovate, to exhibit creativity. As AI amplifies, augments or replaces our own intelligence, it will inevitably evolve into our most powerful and widely applicable technology. Indeed, artificial intelligence may ultimately prove to be one of the most effective tools we have as we look to recover from the crisis unleashed by the coronavirus.

What's more, it's a good bet that artificial intelligence will rise to dominance far faster than was the case with electricity. The reason is that much of the infrastructure required to deploy AI—including computers, the internet, mobile data services and especially the massive cloud computing facilities maintained by companies like Amazon, Microsoft and Google—is already in place. Imagine how rapidly electrification might have occurred if most power plants and transmission lines had already been built at the time Edison invented the light bulb. Artificial intelligence is poised to reshape our world—and it may happen much sooner than we expect it.

AN "ELECTRICITY OF INTELLIGENCE"

The analogy to electricity is apt in that it conveys the sense that artificial intelligence will be ubiquitous and universally accessible and that it will ultimately touch and transform nearly every aspect of our civilization. There are, however, critical differences between the two technologies. Electricity is a fungible commodity that is static over both place and time. Regardless of your location or the company that supplies electric power, the resource you access through the electrical grid is essentially the same. Likewise, the electric power on offer today is little changed from what was available in 1950. Artificial intelligence, in contrast, is far less homogeneous and vastly more dynamic. AI will supply myriad and constantly changing capabilities and applications and may vary dramatically based on who exactly is supplying the technology. And as we will see in Chapter 5, artificial intelligence will relentlessly continue to advance, gaining capability and pushing ever closer to human-level intelligence, and perhaps someday beyond.

While electricity provides the power that enables the operation of other innovations, AI directly delivers intelligence—including the ability to solve problems, to make decisions and in all likelihood to someday reason, innovate and conceive new ideas. Electricity might power a labor-saving machine, but AI is itself a labor-saving technology, and as it scales across our economy, it will have enormous implications for the human workforce and the structure of businesses and organizations.

As artificial intelligence continues to evolve into a universal utility, it will shape the future in much the same way that electricity provided a foundation for modern civilization. Just as buildings and other infrastructure are designed and constructed to take advantage of the existing electrical grid, future infrastructure will be designed from the ground up to leverage the power of AI. And this idea will extend beyond physical structures to transform the design of nearly every aspect of our economy and society. New businesses or organizations will be set up to take advantage of AI from their inception; artificial intelligence will become a critical component of every future business model. Our political and social institutions will likewise evolve to incorporate and rely on this ubiquitous new utility.

The upshot of all this is that AI will ultimately achieve the reach of electricity, but it will never have the same stability or predictability. It will always remain a vastly more dynamic and disruptive force with the potential to upend nearly anything it touches. Intelligence is, after all, the ultimate resource—it is the fundamental capability that underlies everything human beings have ever created. It is difficult to imagine a development more consequential than the transformation of that resource into a universally accessible and affordable utility.

THE EMERGING HARDWARE
AND SOFTWARE AI INFRASTRUCTURE

Like any utility, artificial intelligence will require an enabling infrastructure, a network of conduits that allows the technology to be universally delivered. This begins, of course, with the vast computing infrastructure already in place, including hundreds of millions of laptop and desktop computers, as well as servers in massive data centers, and a rapidly expanding universe of ever more capable mobile devices. The effectiveness of this distributed computing platform as a delivery vehicle for AI is being dramatically improved by the introduction of a range of hardware and software specifically designed to optimize deep neural networks.

This evolution began with the discovery that special graphics microprocessors, used primarily to make fast-action video games possible, were a powerful accelerant for deep learning applications. Graphics processing units, or GPUs, were originally designed to turbocharge the computations required to almost instantaneously render high-resolution graphics. Beginning in the 1990s, these specialized computer chips were especially important in high-end video game consoles, such as the Sony PlayStation and Microsoft Xbox. GPUs are optimized to rapidly perform a vast number of calculations in parallel. While the central processing chip that powers your laptop computer might have two, or perhaps four, computational "cores," a contemporary high-end GPU would likely have thousands of specialized cores, all of which can crunch numbers at high speed simultaneously. Once researchers discovered that the calculations required by deep learning applications were broadly similar to those needed to render graphics, they began to turn en masse to GPUs, which rapidly evolved into the primary hardware platform for artificial intelligence.

Indeed, this transition was a key enabler of the deep learning revolution that took hold beginning in 2012. In September of that year, a team of AI researchers from the University of Toronto put deep learning on the technology industry's radar by prevailing at the ImageNet Large Scale Visual Recognition

Challenge, an important annual event focused on machine vision. Without relying on GPU chips to accelerate their deep neural network, it's doubtful that the winning team's entry would have performed well enough to win the contest. We'll delve further into the history of deep learning in Chapter 4.

The University of Toronto's team used GPUs manufactured by NVIDIA, a company founded in 1993 whose business focused exclusively on designing and manufacturing state-of-the-art graphics chips. In the wake of the 2012 ImageNet competition and the ensuing widespread recognition of the powerful synergy between deep learning and GPUs, the company's trajectory shifted dramatically, transforming it into one of the most prominent technology companies associated with the rise of artificial intelligence. Evidence of the deep learning revolution manifested directly in the company's market value: between January 2012 and January 2020 NVIDIA's shares soared by more than 1,500 percent.

As deep learning projects migrated to GPUs, AI researchers at the leading tech companies began to develop software tools designed to jump-start the implementation of deep neural networks. Google, Facebook and Baidu all released open-source software that was free for others to download, use and update, geared toward deep learning. The most prominent and widely used platform is Google's TensorFlow, released in 2015. TensorFlow is a comprehensive software platform for deep learning, providing both researchers and engineers working on practical applications with optimized code to implement deep neural networks, as well as a range of tools to make the development of specific applications more efficient. Packages like TensorFlow and PyTorch, a competing development platform from Facebook, free researchers from writing and testing software code to deal with arcane details and allow them to instead take a higher-level perspective as they build systems.

As the deep learning revolution progressed, NVIDIA and a number of competing companies moved to develop even more powerful microprocessor chips that were specifically optimized for deep learning. Intel, IBM, Apple and Tesla all now design computer chips with circuitry designed to accelerate the computations required by deep neural networks. Deep learning chips are finding their way into a myriad of applications including smartphones, self-driving cars and robots as well as high-end computer servers. The result is an ever-expanding network of devices designed from the ground up to deliver artificial intelligence. Google announced its own custom chip, called a Tensor Processing Unit or TPU, in 2016. TPUs are specifically designed to optimize deep learning applications built with the company's TensorFlow software platform. Initially, Google deployed the new chips in its own data centers, but beginning in 2018, TPUs were incorporated into the servers that power the company's cloud computing facilities, making state-of-the-art deep learning capability easily accessible to clients who utilize its cloud computing service—a development that would contribute to the dominance of what has become the single most important conduit for the widespread distribution of artificial intelligence capability.

The competition between the established makers of microprocessor chips, as well as a new crop of startups, for a share of the rapidly growing artificial intelligence market has injected a vibrant burst of innovation and energy into the industry. Some researchers are pushing chip designs in entirely new directions. The specialized deep learning chips that evolved from GPUs are optimized to speed up the demanding mathematical calculations performed by software that implements deep neural networks. A new class of chip comes much closer to mimicking the brain, largely dispensing with the resource-hungry software layer and implementing neural systems in hardware.

These emerging "neuromorphic" chip designs instantiate hardware versions of neurons directly in silicon. IBM and Intel have both made significant investments in research into neuromorphic computing. Intel's experimental Loihi chips, for example, implement 130,000 hardware neurons, each of which can connect to thousands of others.[1] One of the most important advantages of eliminating the requirement for software computation at massive scale is power efficiency. The human brain, with capability far beyond any existing computer, consumes only about twenty watts—substantially less than an average incandescent light bulb. Deep learning systems running on GPUs, in contrast, require vast amounts of electricity, and as we'll see in Chapter 5, scaling these systems to consume ever more resources is likely unsustainable. Neuromorphic chips, with designs directly inspired by the brain's neural network, are far less power hungry. Intel claims that its Loihi architecture is up to 10,000 times more energy efficient than traditional microprocessor chips in some applications. Once designs like Loihi enter commercial production, they are likely to be quickly incorporated into mobile devices and other applications where power efficiency is a top concern. Some AI experts go much further and predict that neuromorphic chips represent the future of artificial intelligence. One analysis from the research firm Gartner, for example, projects that neuromorphic designs will largely displace GPUs as the primary hardware platform for AI by 2025.[2]

CLOUD COMPUTING AS THE PRIMARY INFRASTRUCTURE FOR ARTIFICIAL INTELLIGENCE

Today's cloud computing industry got its start in 2006 with the launch of Amazon Web Services, or AWS. Amazon's strategy was to leverage its expertise in building and managing the massive data centers that powered its online shopping service by selling flexible access to computing resources hosted in similar facilities to a wide range of clients. As of 2018, Amazon Web Services operated more than one hundred data centers located in nine different countries throughout the world.[3] The growth of the cloud services provided by Amazon and its competitors has been staggering. According to one recent study, a full ninety-four percent of organizations, ranging from multinational corporations to small- and medium-sized businesses, now utilize cloud computing.[4] By 2016, AWS was growing so fast that the new computing resources that Amazon had to add to its system *every day* were roughly equivalent to everything the company had in place at the end of 2005.[5]

Before the advent of cloud providers, businesses and organizations needed to purchase and maintain their own computer servers and software and to employ a team of highly paid technologists to continuously maintain and upgrade the systems. With cloud computing, much of this is instead outsourced to providers like Amazon, who are able to achieve a ruthless level of efficiency by taking advantage of economies of scale. Facilities that host cloud computing servers are typically massive, encompassing hundreds of thousands of square feet in structures costing upward of a billion dollars and hosting more than 50,000 powerful servers. Cloud computing resources are often provided as an on-demand service in which clients utilize and pay for only the computing power, storage and software applications required at any given time.

Though the facilities that host cloud servers are physically of massive scale, they rely so heavily on automation that they often employ remarkably few people. Sophisticated algorithms deployed to manage nearly everything that happens inside these structures allow for a level of precision that would be impossible under direct human control. Even factors such as the vast amounts of electrical power consumed by the facilities and the need to provide cooling to offset the massive amounts of heat generated by tens of thousands of servers are often optimized from moment to moment. Indeed, one of the first practical applications of DeepMind's AI research was a deep learning system that could optimize the cooling systems in Google's own data centers. DeepMind claims their neural network, which was trained on a trove of data collected from sensors distributed throughout Google's hosting facilities, has been able to cut the energy used for cooling by up to forty percent.[6] Algorithmic control has produced real benefits. A study published in February 2020 found that "while the amount of computing done in data centers increased by about 550 percent between 2010 and 2018, the amount of energy consumed by data centers only grew by six percent during the same time period."[7] All this automation, of course, has an impact on employment. The transition to cloud computing and the resulting evaporation of huge numbers of jobs held by technical experts who once managed computing resources maintained by thousands of individual organizations likely made a significant contribution to the dampening down of the technology jobs boom that occurred in the late 1990s.

The cloud computing business model is highly lucrative and competition among the major providers is intense. AWS is far and away the most profitable part of Amazon's operations, with margins far in excess of the company's e-commerce activities. In 2019, revenue from AWS grew thirty-seven percent to $8.2 billion, and the cloud service accounted for about thirteen percent of the company's total earnings.[8] Amazon's AWS remains the dominant force, with roughly a third of the overall cloud computing market. Microsoft's Azure service, established in 2008, and Google Cloud Platform, launched in 2010, also have significant shares of the market. IBM, the Chinese e-commerce giant Alibaba and Oracle are likewise important players.

Governments as well as businesses are now highly dependent on cloud computing. In 2019, the complexities and partisan tensions inherent in this reliance were thrust into the limelight when the Pentagon's JEDI project turned into a political football. JEDI, an acronym for the Joint Enterprise Defense Infrastructure project, is a ten-year, $10 billion contract to host massive quantities of data and to provide software and artificial intelligence capability to the U.S. Department of Defense. The first kerfuffle occurred at Google, when its employees—who tend to have views positioned pretty far left on the political spectrum—objected to the company's plans to bid for the defense-related contract. Employee protests eventually led Google to take itself out of the running, and the company withdrew just three days before bids on the JEDI contract were due.[9]

Eventually, the Pentagon awarded the project to Microsoft Azure, but Amazon, which because of its leadership in the sector was seen as the most likely winner, immediately claimed the decision was politically motivated. Amazon filed a lawsuit in December 2019 claiming the decision was improperly biased because of President Donald Trump's overt animosity toward Amazon CEO Jeff Bezos. Bezos also owns the *Washington Post*, which has been highly critical of the Trump Administration. In February 2020, a federal judge issued an injunction temporarily blocking award of the contract to Microsoft.[10] A month later, the Department of Defense said it would reconsider its decision.[11]

All this offers a pretty vivid illustration of just how ferocious, and in some cases politically fraught, the battle for the cloud computing market is certain to be going forward. And at the very center of that competitive dynamic stands the artificial intelligence capability that has become an ever more critical component of the services offered by the leading cloud computing providers. The commercial importance of deep learning was initially demonstrated through the tech giants' efforts to deliver their own leading-edge consumer and business services. Neural networks running on specialized hardware within internal data centers, for example, power Amazon's Alexa, Apple's Siri and Google's Assistant and Translate services. From this starting point, deep learning capability has now fully migrated into the cloud services offered by these companies, and it has emerged as one of the most important parameters along which the providers differentiate themselves. Google, for example, has leveraged the popularity of its TensorFlow platform by offering its cloud clients direct access to powerful hardware built from its TPU chips. Amazon, in turn, provides deep learning capability utilizing the latest GPUs and lets its clients run applications created using TensorFlow or a variety of other machine learning platforms. Indeed, Amazon claims that eighty-five percent of cloud AI applications developed with Google's TensorFlow actually run on its own AWS service.[12]

Among the major cloud companies, there is a relentless drive to offer more flexibility and better tools and to rapidly respond to any advantage gained by a competitor. In one recent example of innovation at the technical frontier, Intel made an experimental neuromorphic computing system available via the cloud in March 2020. The system, built from 768 of Intel's brain-like Loihi chips, contains one hundred million hardware neurons—roughly equivalent to the brain of a small mammal.[13] If such architectures prove effective, a neuromorphic battle between the major cloud providers is certain to unfold in short order. As the companies strive to one-up each other and capture a larger share of the ever-increasing market for AI-oriented computing resources, the result has been the emergence of a cloud ecosphere built from the ground up to deliver artificial intelligence.

Microsoft's 2019 billion-dollar investment in the AI research company OpenAI—which along with Google's DeepMind is a leader in pushing the frontiers of deep learning—offers a case study in the natural synergy between cloud computing and artificial intelligence. OpenAI will be able to leverage massive computational resources hosted by Microsoft's Azure service— something that is essential given its focus on building ever larger neural networks. Only cloud computing can deliver compute power on the scale that OpenAI requires for its research. Microsoft, in turn, will gain access to practical innovations that are spawned by OpenAI's ongoing quest for artificial general intelligence. This will likely result in applications and capabilities that can be integrated into Azure's cloud services. Perhaps just as importantly, the Azure brand will benefit from an association with one of the world's leading AI research organizations and better position Microsoft to compete with Google, which enjoys a strong reputation for AI leadership, in part because of its ownership of DeepMind.[14]

This synergy extends far beyond this single example. Virtually every important initiative in AI, ranging from university research labs to AI startups to practical machine learning applications being developed in large corporations, increasingly relies on this nearly universal resource. Cloud computing is arguably the single most important enabler of the evolution of artificial intelligence into a utility that is poised to someday become as ubiquitous as electricity. Fei-Fei Li, the architect of the ImageNet dataset and competition that became a catalyst for the deep learning revolution, took a sabbatical from her current position at Stanford to act as Google Cloud's chief science officer from 2016 to 2018. She puts it this way: "If you think about disseminating technology like AI, the best and biggest platform is a cloud because there's no other computing on any platform which humanity has invented that reaches as many people. Google Cloud alone, at any moment, is empowering, helping, or serving billions of people."[15]

TOOLS, TRAINING AND THE DEMOCRATIZATION OF AI

The evolution of cloud-based artificial intelligence into a general utility is being accelerated by the emergence of new tools that make the technology accessible to a wide range of people who don't necessarily have highly technical backgrounds. Platforms such as TensorFlow and PyTorch do make it easier to build deep learning systems, but they are still by and large used by highly trained experts, often with PhDs in computer science. New tools such as Google's AutoML, introduced in January 2018, largely automate many of the technical details and lower entry barriers substantially, giving far more people the opportunity to utilize deep learning to solve practical problems. AutoML essentially amounts to deploying artificial intelligence to create more artificial intelligence and is part of a trend that Fei-Fei Li calls "the democratization of AI."

As always, competition between the cloud providers is a powerful driver of innovation, and Amazon's deep learning tools for the AWS platform are likewise becoming easier to use. Along with the development tools, all the cloud services offer pre-built deep learning components that are ready to be used out of the box and incorporated into applications. Amazon, for example, offers packages for speech recognition and natural language processing and a "recommendation engine" that can make suggestions in the same way that online shoppers or movie watchers are shown alternatives that are likely to be of interest.[16] The most controversial example of this kind of prepackaged capability is AWS's Rekognition service, which makes it easy for developers to deploy facial recognition technology. Amazon has come under fire for making Rekognition available to law enforcement agencies, given that some tests have suggested the package can be susceptible to racial or gender bias—an ethical issue we will examine more closely in Chapters 7 and 8.[17]

A second crucial trend is the arrival of online training platforms that allow anyone with sufficient initiative and mathematical ability to achieve basic competence in deep learning. Examples include deeplearning.ai, which is offered through the online education platform Coursera, and fast.ai, which offers completely free online courses and software tools that make deep learning more accessible.[18] In an employment landscape where the path to the upper middle class nearly always requires formal credentials obtained through massive investments of time and money, becoming a deep learning practitioner—at least in the current environment, where demand for workers far outstrips supply—is a rare exception. Anyone who can successfully complete the online course work and demonstrate proficiency working with deep neural networks has a good shot at launching a lucrative and rewarding career.

As both training and tools get better and as more developers and entrepreneurs begin to deploy AI applications, we are likely to see a kind of Cambrian explosion as the technology is applied in a myriad of different ways. Something similar has occurred on other major computing platforms. I was running a small software company in Silicon Valley in the 1990s when Microsoft Windows emerged as the dominant platform for personal computers. Initially, Windows application development was a highly technical affair involving the C programming language and thousand-page manuals packed with arcane details. The emergence of easier-to-use tools, including highly accessible development environments like Microsoft's Visual Basic, dramatically expanded the number of people who could engage in Windows programming and soon led to an explosion of applications. Mobile computing has followed a similar trajectory, and both Apple's App Store and the Android Play Store now offer a seemingly infinite number of apps to address nearly any conceivable need. The same sort of explosion is likely coming to artificial intelligence, and more specifically to deep learning. The emergence of AI as the new electricity will, for the foreseeable future, be driven by an ever-expanding spectrum of specific applications rather than any more general machine intelligence.

AN INTERCONNECTED WORLD AND THE "INTERNET OF THINGS"

The final piece of the "artificial intelligence as the new electricity" puzzle is vastly improved connectivity. The most important driver of this is likely to be the roll out of fifth-generation wireless service (or 5G) in the coming years. 5G is expected to boost mobile data speeds by at least a factor of ten—and perhaps as much as one hundred—while dramatically increasing network capacity so that bottlenecks are largely eliminated.[19] This will lead inevitably to a far more interconnected world where communication happens almost instantaneously. We can imagine that virtually everything—including devices, appliances, vehicles, industrial machinery and a great many elements of our physical infrastructure—will all be interconnected and often monitored and controlled by smart algorithms running in the cloud. This vision of the future has been dubbed the "Internet of Things" and is poised to usher in a world where, for example, sensors in your refrigerator or elsewhere in your kitchen detect that you're running low on a particular item and then relay that information to an algorithm that alerts you or perhaps even automatically places the necessary online order. If the refrigerator isn't running optimally, another algorithm will often be able to accomplish an automated or remote resolution. A part that is about to fail will be identified and flagged for replacement. Scaling this model across our entire economy and society is likely to deliver enormous efficiency gains as machines, systems and infrastructure automatically diagnose, and often fix, problems as they arise. The Internet of Things will, in many ways, be like unleashing the algorithms that currently operate cloud data centers with a superhuman level of efficiency to run the wider world. All this will, however, also bring with it some very real risks, especially in the areas of security and privacy, and we will focus on these critical issues in Chapter 8.

This ever more connected world will evolve into a powerful platform for the delivery of artificial intelligence. For the foreseeable future, the most important AI applications will be centered in the cloud. However, over time machine intelligence will gradually become more distributed. Devices, machines and infrastructure will become smarter and smarter as they incorporate the latest specialized AI chips. This is where innovations like neuromorphic computing are likely to have a big impact. The upshot of all of this is a powerful new utility carrying the ability to deliver machine intelligence on demand virtually everywhere.

THE VALUE IS IN THE DATA

As the major cloud providers compete on the basis of both price and the capability of their technology, the cost of accessing the hardware and software that enables artificial intelligence seems certain to fall. At the same time, the AI services available through the cloud will be continuously upgraded as the tech giants strive to gain a competitive advantage by incorporating the latest innovations generated by researchers working at the field's frontier. As all this progresses, even the most advanced AI technologies will become increasingly commoditized and available at little or no cost beyond what cloud computing clients pay to host their data. Indeed, there is evidence of this already. Companies like Google, Facebook and Baidu have all released their deep learning software in open-source form; in other words, they give it away for free. This is also true of the most advanced research conducted by organizations like DeepMind and OpenAI. Both publish openly in leading scientific journals and make the details of their deep learning systems available to everyone.

There is one thing, however, that no company gives away for free: its data. This means that the powerful synergy between AI technology and the vast quantities of data it consumes will inevitably be skewed in one direction. Nearly all the value generated will be captured by whoever owns the data. This widely recognized reality often leads to the assumption that the tech giants will completely dominate any sphere that intersects with big data or artificial intelligence. However, this overlooks the fact that data ownership is clearly verticalized by industry and economic sector. Companies like Google, Facebook and Amazon do, of course, control unimaginable troves of data. However, it is generally limited to areas like web search, social media interactions and online shopping transactions. In these arenas, the established companies are likely to remain dominant, but far more data of completely different kinds resides across the economy and society, under the control of governments, organizations and businesses in other industries.

It's often said data is the new oil. If we embrace this analogy, then it's fair to say that the tech companies in many ways fulfill a role similar to that of perhaps Halliburton, offering the technology and know-how required to extract value from the resource. The tech giants do, of course, also control huge data reserves of their own, but still the lion's share of this ever-expanding global data resource lies in the hands of others. Businesses like health insurers, hospital networks and, of course, government-managed national health services control data of immense value. To be sure, they will employ the latest AI technology developed by the big technology companies and delivered through the cloud, but they will largely retain the value extracted from their data. The same will be true of the massive amounts of data generated by financial transactions, travel bookings, online reviews, customer movements within physical retail stores and operational data generated by a myriad of sensors built into vehicles and industrial machinery. In each case, the ubiquitous new utility of machine intelligence will be applied to specific types of data owned by entities distributed across the economy.

One important implication of this is that much of the value derived from the application of artificial intelligence will be captured by entities beyond the obvious candidates within the technology sector. The enormous benefits from leveraging AI will be distributed widely. Again, an analogy to electricity is useful here. Who generates the most value from electricity? Is it electric utilities? Or the nuclear power industry? No, it's companies like Google and Facebook that consume massive amounts of electricity and have discovered ways to transform this ubiquitous commodity into fantastic value. The analogy is, of course, not perfect, and without doubt, immense value and power will reside in those companies that innovate on the frontier of artificial intelligence and deliver this ever-improving resource. But most of the benefits that arise from the application of AI—especially as it increasingly resembles a commoditized utility—are likely to accrue elsewhere.

While the value created by artificial intelligence will be distributed widely across economic sectors, the reverse may well turn out to be true within a given industry. Companies that are on the frontier when it comes to leveraging AI within their business models are likely to have a substantial first-mover advantage. That could well lead to winner-take-all scenarios as businesses with especially effective big data and artificial intelligence strategies gain a significant competitive advantage. Because data is so central to the effective application of AI, the first step toward an AI strategy is nearly always a successful data strategy. This means that it is crucial for businesses and organizations to focus on building efficient data collection and management systems as a prelude to deploying AI. In some cases, this will require addressing important ethical considerations, for example, around issues of privacy concerning employees and customers. However, those organizations that fail to move aggressively are likely to be left behind. We are rapidly moving toward a reality where any business, government or organization that leaves artificial intelligence on the table is engaging in a misstep of such magnitude that it could reasonably be compared to disconnecting from the electrical grid.

AS ARTIFICIAL INTELLIGENCE evolves into a truly universal utility, reaching into every business, organization and household, it will inevitably transform both our economy and our society. This is a story that will play out over the course of years and decades, and the impact won't be uniform. In some areas AI is likely to be transformative in the next few years, while in other cases the disruption will take much longer to arrive. The next chapter looks at some of the practical implications of artificial intelligence as a systemic technology, attempts to separate hype from reality and delves into the intersection between this rapidly advancing technology and the pandemic that has completely upended all our lives.

CHAPTER 3

BEYOND HYPE: A REALIST'S VIEW OF ARTIFICIAL INTELLIGENCE AS A UTILITY

ON APRIL 22, 2019, TESLA HELD AN EVENT IT DUBBED "AUTONOMY Day." Intended to highlight the autonomous driving technology the company builds into every Tesla, the event featured presentations by CEO Elon Musk and other top executives and engineers. At the event, Musk said, "I feel very confident predicting autonomous robotaxis for Tesla next year." He went on to suggest that Tesla would have a million such cars operating on public roads by the end of 2020.[1] By "robotaxis," Musk meant genuine self-driving cars, capable of operating with no one inside and able to pick up passengers and deliver them to random locations. In other words, a truly robotic version of Uber or Lyft.

This was an astonishing prediction: far out of line with the expectations of virtually every other expert I have talked to. A few days later, I appeared on Bloomberg TV and said that I was "astounded by" Musk's prediction and that I thought it was "extraordinarily optimistic and perhaps even a bit reckless." I said this because such an aggressive prediction would almost certainly result in market pressure on Tesla to deliver, and this combined with the company's ability to provide new features to Tesla owners via software download could be very dangerous if unproven software that purports to deliver fully autonomous capability is suddenly put into the hands of drivers. While it may be fine for a company to have its customers test early versions of a new video game or social media application, this is not a responsible strategy for software that could clearly result in injury or death.* [2] Indeed, there have already been fatal accidents involving Tesla's autopilot feature, which steers, accelerates or brakes the car to stay within its lane but still requires driver supervision. In addition, it seemed clear to me that even in the unlikely event that the company was able to perfect the technology within a year or so, it would take much longer to adequately test the cars and obtain regulatory approval. So, a million operating Tesla robotaxis by the end of 2020 was just not going to happen. Even a single truly autonomous car operating on public roads within that time frame would be astonishing.

Much of the Autonomy Day event was devoted to a discussion of a custom new self-driving microprocessor chip being developed by Tesla. Previously the company has used chips optimized for deep neural networks manufactured by NVIDIA.

---

* In October 2020, Tesla did, in fact, release an early version of what it calls a "Full Self-Driving Package." The software was made available to a limited number of Tesla owners via download, with plans to expand availability over the ensuing months. The software provides features such as automatic parking and a limited ability to navigate city streets but is currently nothing close to what could reasonably be called "full self-driving." Tesla has promised to upgrade the package and has announced future price increases in order to incentivize owners to purchase an early version. The National Highway Traffic Safety Administration took note and declared that it would "monitor the new technology closely" and that it "will not hesitate to take action to protect the public against unreasonable risks to safety." (See endnote 2, Chapter 3.)

Tesla claimed that its new chip offered unprecedented power, but executives at NVIDIA quickly pushed back, pointing out that the latest versions of their AI chips were equivalent to or even faster than the product under development at Tesla.[3]

Nonetheless, as I watched Autonomy Day unfold, it became clear to me that Tesla does indeed have a striking competitive advantage—something that ultimately could allow it to outpace its competitors and be the first company to deploy fully autonomous self-driving cars. This advantage is not a special computer chip, or even an algorithm. Rather—as is so often the case in the field of artificial intelligence—the advantage lies in the data that Tesla controls. Every Tesla is equipped with eight cameras that operate continuously, capturing images from the road and the environment around the car. Computers onboard the cars are able to evaluate these images, determine which ones are likely of interest to the company and then automatically upload these in a compressed format to Tesla's network. Over 400,000 of these camera-equipped cars are driving on roads throughout the world, and that number is increasing rapidly. In other words, Tesla has access to a truly massive trove of real-world photographic data that none of its competitors can come close to matching.

Tesla's director of AI, Andrej Karpathy, described how the company can request specific kinds of images from its "fleet" of camera-equipped cars. For example, if Tesla's engineers want to train its autonomous driving system to handle situations where roads are being worked on, it can summon thousands of real-world images of construction sites and then use those images to train its self-driving software in computer simulations. While all self-driving car initiatives make heavy use of simulation, Tesla's ability to incorporate massive amounts of real-world data is a potentially disruptive advantage. As it is often said, truth is stranger than fiction, and no engineer could design a simulation that would come close to replicating the detailed, and often weird, reality captured by the cameras on Tesla's ever-expanding fleet.

This example illustrates how news about ongoing progress in artificial intelligence is often a kind of obtuse mixture of hype and sensationalism woven into a narrative that also conveys important information. As I have stated, artificial intelligence is destined to become a ubiquitous utility that will ultimately touch virtually everything. However, progress will not be homogeneous: some technical problems are much more difficult to solve than others. In particular, some of the highest profile and most-hyped applications of artificial intelligence are likely to underperform relative to our expectations, while dramatic progress in other, often less visible, arenas will take us by surprise. This chapter will present some examples and guidelines that offer insight into the areas where I think AI is likely to be disruptive in the relatively near term—and where it is likely to take much longer.

DELIVERY OF YOUR HOME ROBOT HAS BEEN DELAYED

The promise of a personal robot for the home—a machine capable of cleaning the house and doing the laundry even as it stands ready to serve as a tireless butler—has captured our collective imagination since some of the first science fiction authors began to speculate about the future. What are the prospects for such a machine? For the moment, set aside the truly advanced fictional examples most of us are familiar with—Rosie the Robot from *The Jetsons* or a humanoid machine like C-3PO from *Star Wars*—and consider something far less ambitious: a functional robot that would have a useful, even if somewhat limited, ability to tidy up a room, perform a variety of basic household cleaning tasks, and perhaps even, on command, bring us a beer from the refrigerator. How soon should we expect to see a reasonably affordable personal robot that we might find so useful, perhaps even indispensable, that masses of value-conscious consumers would be willing to pay for it?

The unfortunate reality is that such a machine probably lies quite far in the future. Indeed, the problem with the personal robots that have been attempted so far is that they simply can't do very much. The minimal requirements for a truly useful machine, including the visual perception, mobility and dexterity needed to function in an unpredictable environment like a home, are among the most daunting challenges in robotics. So far, the companies that have attempted to bring consumer robots to market haven't really even begun to overcome these challenges. Instead, they've produced machines so limited that, for most people, the value proposition is quite questionable.

One example that illustrates these challenges is Jibo, a machine marketed as the first "social robot." Conceived by MIT's Cynthia Breazeal, one of the world's top experts on robots with the ability to engage with people on a social and emotional level, and introduced in the fall of 2017, Jibo is a plastic table-top robot about twelve inches tall. Jibo has no arms, legs or wheels, but it does have the ability to tilt and swivel its head and create at least the illusion of a human-like connection as it communicates with its owner. The robot is able to engage in rudimentary conversations and do a number of practical things that center around information retrieval; it can look up things on the internet, get weather and traffic reports, play music and so forth. In other words, Jibo offers a set of capabilities that are broadly similar to Amazon's Alexa-powered Echo smart speakers. The Echo, of course, can't move at all, but backed by Amazon's massive cloud computing infrastructure and far larger team of highly paid AI developers, it's information retrieval and natural language capabilities are likely stronger—and certain to become more so over time. Jibo's biggest problem was its price of around $900. It turned out that, though the robot's ability to mimic human head gestures and dance along with the music it played was cute and endearing, this capability simply wasn't worth the extra $800 or so to most consumers. The startup company that made Jibo, after burning through a reported $70 million in venture funding, shut down in November 2018.[4]

Amazon is reportedly working on its own home robot. Codenamed Vesta, the machine has been described as a kind of "Echo on wheels," with the ability to navigate around your home and arrive on command.[5] Still, I have seen no reports that Amazon plans to add an arm to the robot or that it will have any ability to physically manipulate its environment. In the absence of such features, one is again left to wonder about the value proposition. Given that the cheapest versions of the Echo are priced at $50 or less, why would it be better to purchase an expensive mobile (but likely quite slow) Echo, rather than simply place cheap stationary versions throughout your home? These are the kinds of questions that haunt the personal robotics industry, and it's unclear that even Amazon will be able to launch a successful commercial product anytime soon.

To get some perspective into how high the hurdle really is for a truly functional home robot, think about just one prospective task: the ability to retrieve a beer from the refrigerator. Assuming there are no major obstacles, such as a staircase or a closed door, getting to the refrigerator is likely the easy part. The technology for robots to navigate within known environments is already in place, as demonstrated, for example, by the Roomba robot vacuum cleaners.

Once the robot arrives, however, it needs to open the refrigerator door. Try this yourself and notice how much force is required. But it's not just a question of brute strength. You can easily open the door because you likely weigh more than a hundred pounds. Consider the physics of the situation. Any robot that can succeed in opening the refrigerator door is not some plastic toy. It is not an Amazon Echo on wheels. A machine that will do anything other than simply tip over needs to be quite heavy, and to manipulate an environment designed for people, it needs to come reasonably close to human proportions. This machine is going to be expensive. Even if we can figure out a cheap way to provide the necessary counterweight—perhaps filling a plastic robot with water, for example—the required weight still implies the need for a powerful motor and heavy-duty wheels to propel the robot.

Once the door is open, the robot needs to locate the beer. What if it's hidden behind the takeout food containers left over from last night's dinner? What if the cans of beer are bundled in plastic six-pack rings? Could a robot successfully remove a can? Think about how the mechanics of doing so might be completely different depending on how many cans of beer are left. Is it a full six pack or one lonely can still attached to the plastic? A robot that could do this simple thing would need to be extraordinarily dexterous and would likely need to have two very expensive robotic arms—not just one.

Of course, it's easy to imagine ways to get around some of these problems. Maybe the beer needs to be placed in exactly the right location within the refrigerator. Forget six-pack rings. The cans must be removed from any packaging, and maybe each must be fitted with an RFID tag so the robot doesn't have to rely solely on visual perception to find the beer. Perhaps, someday, beer will come in some sort of futuristic packaging that is specifically designed to make it easy for robots to retrieve it. But for now, all these requirements would add to your inconvenience and therefore diminish your enthusiasm for making a major outlay to acquire such a robot.

And make no mistake, any genuinely functional home robot will require a very substantial financial investment. Components like electric motors, robotic arms and the various kinds of sensors required to give a robot visual perception, spatial orientation and tactile feedback are not subject to the Moore's Law–driven cost deflation that has characterized the semiconductor industry and made computing power ever more affordable. The essential problem for a home robot is that, in order to offer real value to consumers, it needs to at least approach our own manipulative capabilities. And human beings, it turns out, are astonishingly effective biological robots.

Imagine two objects on a table in front of you. On the left is a solid steel bearing, three inches in diameter and weighing in at about four pounds. To the right is an egg. You can easily pick up either of these objects. Think about the forces that the muscles in your hand need to apply as you grip each object and then begin to lift it. Consider the consequences if you somehow confused the objects and applied the wrong set of forces. Even if you were blindfolded, it's a good bet that you would succeed in safely picking up both objects on the basis of tactile feedback alone. The motors and sensors required to replicate that ability in a robotic hand would be expensive—even if the controlling software to make it possible were available.

The reality is that, even after decades of work on robotic hands and the algorithms required to animate them, their dexterity is not yet close to human level. Rodney Brooks, one of the world's foremost roboticists and a co-founder of iRobot Corporation, the makers of the Roomba as well as some of the world's most advanced military robots, illustrates this by alluding to the long reach plastic gripper tools that you often see being used to pick up trash:

That really primitive [gripper] can do fantastic manipulation beyond what any robot can currently do, but it's an amazingly primitive piece of plastic junk. . . . That's the clincher: you are doing the manipulation. Often, you'll see videos of a new robot hand that a researcher has designed, and it's a person holding the robot hand and moving it around to do a task. They could do the same task with this little plastic grabber toy; it's the human doing it. If it was that simple, we could attach this grabber toy to the end of a robot arm and have it perform the task—a human can do it with this toy at the end of their arm, why can't a robot? There's something dramatic missing.[6]

Even if a robot tasked with tidying up a home were to achieve the necessary level of dexterity, it still faces the challenge of recognizing the many thousands of different objects it might encounter, and then figuring out what to do with them. Which things should be carefully returned to their appropriate place and what items are trash that should be discarded? What error rate would you be willing to tolerate in an unsupervised robot turned loose in even a single room of your home?

None of this is to say that your home robot will *never* arrive. Significant progress toward overcoming many of these hurdles is already being made. For example, it seems likely that future robots will be able to recognize the objects they encounter by interfacing with the cloud. You can already see a pretty impressive demonstration of this with Google's Lens service, which allows you to point your mobile phone at nearly anything and automatically generate an identification, as well as descriptive information and examples of similar objects.

As the world becomes more connected and the Internet of Things gains traction, sensors of the type used in robots will be deployed widely in a variety of applications, and as demand for these devices grows, the resulting economies of scale should drive down costs. The same is likely to eventually happen with other components as robots increasingly penetrate the commercial sector.

Likewise, researchers are successfully deploying deep learning and other techniques to build more dexterous robotic hands. One of the highest-profile demonstrations came from OpenAI, when in October 2019, it announced that it had created a system consisting of two integrated deep neural networks that enabled a robotic hand to solve a Rubik's Cube.[7] The system was trained using high-speed simulation and succeeded only after the equivalent of about 10,000 years of reinforcement learning. Solving a Rubik's Cube with one hand is not at all easy even for humans. Despite the company's claim that it had achieved something "close to human-level dexterity," it turned out that it wasn't easy for OpenAI's system either: the robotic hand dropped the cube in eight out of ten attempts.[8] Still, initiatives like this represent real progress, and as we will see, in many industrial and commercial environments, improving robotic dexterity will begin to have a significant impact within the next few years. Until the artificial intelligence needed to animate robots in highly unpredictable environments gets much better and the necessary components become dramatically cheaper, however, an affordable and genuinely useful robot for the home is likely to remain out of reach for the foreseeable future.

WAREHOUSES AND FACTORIES— GROUND ZERO FOR THE ROBOTIC REVOLUTION

If both technical limitations and economics dictate that a versatile and productive home robot will likely be a long time coming, exactly the opposite is true in many industrial and commercial settings. Within the enclosed space of a factory or warehouse, it's possible to eliminate, or at least minimize, much of the unpredictability and chaos that rules in the outside world. More often than not, this involves reorganizing the interaction and flow of people, machines and materials within the facility to take advantage of the capabilities of robots, while working around their limitations. The value proposition offered by a strict requirement to unfailingly place the beer—as well as every other item—at precise coordinates within your refrigerator in order to enable reliable robotic retrieval may not seem especially compelling, but in high-volume commercial environments where even a tiny increase in efficiency can result in a massive financial return, the calculation is quite different.

There is no better demonstration of all this than the inner workings of the distribution centers operated by Amazon and other online retailers. Behind the walls of these typically massive facilities, the robot revolution is already well underway and is unquestionably poised to accelerate. Less than a decade ago, warehouses of this kind would have nearly always been animated by hundreds of workers continuously roving the aisles between tall shelves containing thousands of different inventory items. The workers would have been generally divided into two groups: "stowers" tasked with taking newly arrived inventory and storing it at the proper positions on shelves, and "pickers" who traveled to these same locations to retrieve items in order to fulfill customer orders. The activity within the warehouse would have been a continuous mad scramble—perhaps resembling an especially disordered anthill—in which a typical worker might well trek a dozen or more miles over the course of a single shift, hurrying back and forth to random locations and often climbing ladders to reach the uppermost shelves.

Within Amazon's most modern distribution centers, this bustling motion has been transformed almost into a mirror image of itself. It is now the workers who remain stationary, while the inventory shelves speed about, hurrying between destinations riding on the backs of fully autonomous robots. This wholesale reorganization began with Amazon's $775 million acquisition of the warehouse robotics startup Kiva Systems in 2012. The robots, which look somewhat like huge orange hockey pucks and weigh in at over 300 pounds, roam within a fenced-off area designed to eliminate any risk of collisions with human workers and navigate by following barcodes attached to the floor. Operating under algorithmic control, robots deliver shelves laden with inventory to stations occupied by workers, who are then tasked with either stowing items at an available location or retrieving a specific product to fulfill a customer order.

Amazon now operates more than 200,000 of these robots at its distribution centers worldwide. The result has been a three- to four-fold increase in the number of items that can be retrieved by a typical picker over the course of an hour.[9] So far, the robots have not, for the most part, replaced workers. Indeed, employment in Amazon warehouses has grown dramatically— offsetting to some extent the evaporation of jobs in traditional retail settings as online shopping has gained traction. The robots rapidly navigate smooth, obstacle-free floors carrying up to 700 pounds of inventory, while workers stay in place, performing tasks requiring the visual perception and dexterity that, so far at least, is beyond the capability of any robot.[10] This synergy between workers and machines has been instrumental in enabling Amazon to continuously ramp up the level of service it offers its customers. One-day delivery for Amazon's Prime customers, introduced in 2019, for example, would have likely been impossible without this massive investment in robotics. Likewise, automation was likely crucial as Amazon struggled to keep pace with soaring demand as the coronavirus crisis unfolded, even as many of its warehouse workers fell sick.

While this coupling of workers and robots in ways that leverage the relative strengths of each results in undeniable efficiency gains, it also transforms the nature of these jobs in ways that can be both positive and negative. Under the new regime, the exhausting trudge through the warehouse aisles has been replaced with mind-numbing repetition. Workers now stand in place and stow or pick items from the arriving shelves hour after hour. According to one analysis, warehouse injuries at Amazon occur at more than twice the warehouse industry average and have actually increased with the new robotic technology in part because of repetitive motion distress or the strain of retrieving heavy items from higher shelves.[11] As Marc Wulfraat, an industry consultant, told Vox reporter Jason Del Rey, "Walking 12 miles a day on a concrete floor to pick these orders. . . . If you're not 20 years old, you're a broken person at the end of the week. . . . Having a rubber mat, where goods come to you, is three times more productive than the traditional approach and more humane, . . . [but] picking three times faster also implies more wear and tear due to repetitive motion and working faster at lifting and handling products."[12]

The truth is that, within facilities like this, workers are gradually losing their agency and being transformed into what essentially amounts to plug-in biological neural networks that fill the gaps in a largely mechanized process by rendering the capabilities that are—so far—beyond the reach of machine intelligence. One result has been protests at distribution centers in both the United States and Europe, complaining that human beings are being treated like robots and that workers are continuously driven to meet unreasonable expectations under the supervision of ever more demanding algorithms.[13] It seems to me that if these jobs are increasingly perceived as dehumanizing or even dangerous, and as workers are pushed ever closer to their physical and psychological limits, it will inevitably become a rational for their elimination as soon as the necessary technology arrives.

Indeed, within these kinds of enclosed and relatively controlled environments, the ratchet of automation is likely to click forward relentlessly, gradually driving operations to become ever less labor intensive. Amazon is already moving aggressively to automate more aspects of its warehouse operations. A May 2019 report by Reuters journalist Jeffrey Dastin revealed that Amazon has been introducing advanced machines capable of doing the final packaging of products into boxes ready to ship to customers. Because robots still lack the dexterity to reliably pick up highly varied products and place them in boxes, the machines instead almost instantly construct custom-sized boxes around an item as it travels along a conveyor belt. The machines can box about 600 to 700 items per hour—as much as five times the capability of a human worker. Two people who had been involved in the project at Amazon told Dastin that this could ultimately lead to the elimination of about 1,300 jobs in as many as fifty-five warehouses located throughout the United States.[14]

Likewise, Amazon has introduced robots that look like smaller versions of its hockey puck–shaped Kiva robots in the sorting centers where the company distributes packages to trucks headed to various destinations. Rather than carrying shelves of inventory, the smaller robots instead tote a single package to a specific location on the sorting center floor—corresponding to a zip code—where the package then slides into a hole in the floor and is sent on its way to a truck waiting below.[15] All this, of course, offers another vivid example of how an entire work environment can be designed and restructured from the ground up to maximize the powerful, albeit limited, capabilities of robotic automation. As robots advance and become more versatile and adept, these environments are sure to be periodically restructured to take advantage of new possibilities and maximize productivity.

Within warehouses and factories, the automation endgame will unfold once robots finally do approach human-level capability in terms of their ability to grasp and manipulate objects. Beyond this point, the specter of a fully automated warehouse where employment is limited to a relatively small number of workers who supervise and maintain the machines becomes a realistic scenario. Amazon has clearly demonstrated a keen interest in achieving this milestone; the company has organized a number of highly touted annual contests in which engineering teams from universities around the world have competed to build robots that can perform the tasks now undertaken by the workers who pick items from the shelves in its warehouses.[16] While building a robotic hand capable of reliably grasping thousands of different items—all of different sizes, weights, shapes, textures and packaging configurations—has proven to be a daunting challenge, progress along this path is inevitable. Speaking at a conference in June 2019, Amazon CEO Jeff Bezos said, "I think grasping is going to be a solved problem in the next 10 years," in spite of the fact that "it's turned out to be an incredibly difficult problem, probably in part because we're starting to solve it with machine vision, so machine vision did have to come first."[17] In other words, the thousands of people currently engaged as inventory stowers and pickers—a majority of the company's warehouse workforce—are arguably on a pretty clear glide path toward being made redundant within a decade or so.

In all likelihood, however, the impact on jobs will begin to manifest long before then. Once again, the key factor is the controlled, relatively predictable environment within a warehouse. It seems to me that in a setting like this, a robot does not have to be anywhere near perfect in order to add significant value.

Indeed, a robot that can reliably handle fifty percent—or per-haps or even less—of the items inventoried in a typical ware-house could drive massive productivity increases, as long as the robot consistently fails in a predictable way. Amazon has at its disposal enormous streams of data that can be used to predict exactly where a fulfillment robot is likely to succeed and where it might fail. From the instant a customer places an online order, the company knows, of course, exactly what items are involved and should have little trouble predicting whether that order is a good candidate for fully robotic fulfillment or if it instead needs to be routed to a human worker. In other words, Amazon can take advantage of robots with limited capability simply by man-aging the flow of work within its distribution centers.

This ability to reliably predict the outcome of a robotic operation and work around failure is really the bright line be-tween a controlled warehouse-type environment, where robots are likely to thrive in the relatively near future, and the far more chaotic outside world, where the challenges for technologies like self-driving cars are likely to be far more daunting. A ware-house robot that can predictably handle half the items it might encounter can be highly useful. A self-driving car operating on a public road that can reliably navigate ninety-nine percent of the situations it encounters may be worse than useless because that outlying one percent virtually guarantees disaster.

A partially capable fulfillment robot is likely made even more valuable by the fact that Amazon's sales are governed by a long-tail distribution in which a relatively small fraction of the products stocked in a warehouse constitute the lion's share of the items that customers tend to order. A robot with the ability to consistently grasp and manipulate a significant percentage of these popular, high-volume items would be an especially ef-fective way to achieve productivity gains. Of course, no robot will be completely reliable, even when tasked with fulfilling only those orders that it is expected to be able to handle. To deal with relatively rare failures, it's easy to imagine one human worker overseeing the operation of several fulfillment robots and interceding only when a problem occurs. The upshot is that rather than warehouse automation arriving en masse only after truly human-level robotic dexterity is achieved, it's more likely to take place gradually, in a piecemeal evolution, in which each stage of the process may require a significant reorganization of the workflow within warehouses.

BEYOND AMAZON AND
THE QUEST FOR TRULY DEXTEROUS ROBOTS

While Amazon's initiatives in robotics attract a great deal of attention because of the company's size and influence, the story is broadly similar at facilities run by its online competitors as well as a variety of brick-and-mortar retail chains. Grocery stores in both North America and Europe, in particular, are aggressively moving toward automation in distribution centers as a way of becoming more efficient and delving into online sales, partly in anticipation of Amazon's looming disruption of the grocery market in the wake of its acquisition of Whole Foods in June 2017.

One of the leaders in this arena is United Kingdom–based Ocado, which runs its own online grocery service and also markets its warehouse automation technology to supermarket chains worldwide. At the company's distribution center in Andover, England, more than a thousand robots run on rails arranged in an elevated grid structure resembling a massive checkerboard. Up to 250,000 crates—each stocking a particular grocery item—can be stored at locations corresponding to squares on the board. The robots navigate on the rails above, grappling and then pulling crates up into their box-like interiors and carrying them to stations where individual items are retrieved and customer orders packed. The robots run autonomously, communicating and navigating around each other via a mobile data network, and periodically returning to docking stations to have their batteries charged.[18] There are even specialized retrieval robots that come to the rescue in the event that one of the crate-carrying robots malfunctions. The Andover facility is capable of processing about 65,000 online grocery orders, containing three and half million individual items, every week.[19]

As in Amazon's warehouses, the robots focus on the logistics of rapidly moving materials, while the primary role for humans among all this automation is the picking and packing that continues to require human dexterity. The wide variety of items that characterizes a typical grocery list—everything from canned goods to boxed items to fresh produce—presents a particular challenge for robotic manipulation. As technology journalist James Vincent points out, "nothing stumps a robot quite like a bag of oranges." The difficulty is that "the bag moves in too many weird ways, there are no obvious bits to grab hold of, and if you squeeze too hard you end up with orange juice instead."[20] Nonetheless, Ocado is already experimenting with robots that attempt to overcome these challenges. The company is employing robotic picker arms that use suction cups to lift items with suitable surfaces, such as cans, as well as soft rubber robotic hands that someday will be able to effectively grasp more fragile items.

The quest to build a truly dexterous robot has become a major focus of Silicon Valley venture capital firms, and a number of well-funded startup companies have emerged, embracing varied approaches as they innovate at the research frontier. One of the highest-profile startups is Covariant, which was founded in 2017 but emerged from stealth mode only in early 2020. The researchers at Covariant believe that "reinforcement learning"—or essentially learning through trial and error—is the most effective way to progress, and the company claims to be building a system based on a massive deep neural network that it calls "universal AI for robots," which it expects to eventually power a variety of machines that can "see, reason and act in the world around them, completing tasks too complex and varied for traditional programmed robots."[21] The company, founded by researchers from the University of California at Berkeley and OpenAI, has received investments and publicity from some of the brightest lights in deep learning, including Turing Award winners Geoffrey Hinton and Yann LeCun, Google's Jeff Dean and ImageNet founder Fei-Fei Li.[22] In 2019, Covariant defeated nineteen other companies in a competition organized by the Swiss industrial robot maker ABB by demonstrating the only system capable of recognizing and manipulating a variety of items without any need for human intervention.[23] Covariant will be working with ABB as well other major companies to imbue industrial robots deployed in warehouses and factories with intelligence that the company believes will eventually match or exceed human-level perception and dexterity.

Many of the startup companies and university researchers working in this area believe, like Covariant, that a strategy founded on deep neural networks and reinforcement learning is the best way to fuel progress toward more dexterous robots. One notable exception is Vicarious, a small AI company based in the San Francisco Bay Area. Founded in 2010—two years before the 2012 ImageNet competition brought deep learning to the forefront—Vicarious's long-term objective is to achieve human-level or artificial general intelligence. In other words, the company is, in a sense, competing directly with higher-profile and far better funded initiatives like those at DeepMind and OpenAI. We'll delve into the paths being forged by those two companies and the general quest for human-level AI in Chapter 5.

One of Vicarious's major objectives has been to build applications that are more flexible—or as AI researchers would say, less "brittle"—than typical deep learning systems. This kind of adaptability is a critical requirement for any robot expected to handle a wide variety of tasks that are currently undertaken by humans. Vicarious's technical co-founder Dileep George, who has led AI research at the company, believes that building robots with the ability to understand and manipulate their environment is an essential waypoint on the path to achieving more general intelligence, and in early 2020, the company revealed that the development of versatile robots geared toward logistics and manufacturing would be its primary near-term business strategy.

Although it is secretive about the details, Vicarious claims to have developed an innovative machine learning system, inspired by the function of the human brain, that it calls a "recursive cortical network."[24] The company is deploying its system to animate robots that have already been placed in production for its initial customers, which include the logistics division of Pitney Bowes and the cosmetic company Sephora. Vicarious's robots have a remarkable ability to improve at their assigned tasks—getting measurably better within hours of their initial operation.[25] The objective is to create robots capable of not just picking items from inventory shelves or bins, but to go beyond that and design machines with genuinely versatile manipulative ability, including functions such as sorting and packing items, replacing the workers who tend factory machines by feeding parts in and out and performing detailed assembly work. Vicarious has raised at least $150 million in venture funding and is backed by some of most prominent names in Silicon Valley, including Elon Musk, Mark Zuckerberg, Peter Thiel and—as you might have expected—Jeff Bezos.

In parallel with its progress in artificial intelligence, Vicarious is also pursuing an innovative "robots as a service" business model, which may eventually prove to be disruptive across a range of industries. Rather than building or selling its own robots, Vicarious instead acquires industrial robots from companies like ABB, integrates them with its proprietary artificial intelligence software and then rents the robots out to companies in a way that's roughly comparable to the way a temporary employment agency might place human workers. The result is that client companies do not have to make the upfront capital investment and long-term commitment that is normally associated with industrial robots. This directly addresses one of the biggest drawbacks of using robots: the machines are expensive to purchase, install and program, and therefore it takes a long time for the investment to pay for itself. Traditional industrial robots, however, lack the flexibility and adaptability of a human workforce. Anytime the processes within a factory or warehouse change—and this can occur often, sometimes in the space of a few months—a time-consuming and expensive reprogramming of the robots is required. This has been one of the primary factors holding back more widespread deployment of robots in these environments. The robots-as-a-service approach, combined with the ability to rapidly train robots for new tasks, is clear evidence that we are approaching a future where robots will be just as adaptable as human workers. And that is likely to be a game changer in a variety of industries.

Vicarious isn't the only company that has recognized the advantages of this business model. A similar approach is being pursued by the Australian automation technology company Knapp, which is utilizing robots powered by software from Covariant. In January 2020, Knapp executive Peter Puchwein told the *New York Times* that the company's strategy is to consistently price its robots below the cost of employing human workers. For example, "if a company paid $40,000 per year to a worker, Knapp would charge about $30,000." "We just go lower," Puchwein told the *Times*. "That's basically the business model. For the customer, it's not very hard to decide."[26] Added to the lower cost, of course, is the reality that robots don't take vacations, never get sick, are never late to work and in general don't suffer from any of the management issues and inconveniences that arise continuously with human workers.

Even as robots become vastly more dexterous and begin to approach human levels of capability, it likely will be a long time before these machines become affordable enough to be employed as consumer products in homes. But in environments like factories and warehouses, where things are more predictable and where the logic of profitability and efficiency will inevitably shift the balance between workers and machines, the disruption is likely to come far sooner. As we've seen, robots are becoming not just more adept at physical manipulation but also more flexible and adaptable, and this will make them increasingly likely to be deployed even in areas like electronic assembly, where the ability to rapidly shift production to accommodate new products is critical. All this will likely prove to be an important chapter in the story of artificial intelligence's evolution into an electricity-like utility whose tentacles reach into virtually every aspect of the economy.

The implications for employment will eventually be significant, especially since warehouses and distribution centers have been a relative job-creation bright spot in recent years as online shopping has continued to disrupt the traditional retail sector. All of this could have especially stark consequences when it is imposed upon recovery from the unfolding economic downturn. Likewise, to the extent that the coronavirus—or for that matter a lingering fear of the next pandemic—continues to be a factor, robotic production will present attractive solutions to problems that arise around a requirement for social distancing or around workers who fall sick. We will explore the potential impact of artificial intelligence and robotics on jobs and the economy more fully in Chapter 6.

THE COMING AI REVOLUTION IN
TRADITIONAL RETAIL AND FAST FOOD

On December 3, 2019, *Bloomberg* published an article entitled "Robots in Aisle Two," which delved into the rise of artificial intelligence, robotics and automation at brick-and-mortar retailers in the United States. The article, written by business reporter Matthew Boyle, pointed out that the major grocery chains were especially interested in adopting new technology in order to ward off what they perceived as a potentially existential threat from Amazon's looming entry into their market. The stodgy grocery industry, whose last major innovation was the introduction of bar code scanners in the late 1970s, was now urgently experimenting with "shelf-scanning robots, dynamic pricing software, smart carts, mobile-checkout systems and automated mini-warehouses in the back of stores" among other new AI-centric technologies.[27]

Still, an industry insider quoted in the article sounded a moderating note. "You won't see robots in Target anytime soon," the company's CEO said. "The human touch still really matters."[28] Roughly two days before the article appeared on Bloomberg's website, the first documented case of COVID-19 had emerged in Wuhan, China. Over the course of the next few months, all our calculations around the perceived value of a "human touch" were, of course, reset and recalibrated with a speed that would have once been unimaginable. There can be little doubt that in virtually any environment where human workers come into direct contact with customer traffic, the coronavirus crisis is going to significantly accelerate the push toward automation. This will be true not only because of concerns around social distancing and hygiene, but also as a result of the inevitable escalation of a focus on efficiency in the wake of the economic downturn spawned by the virus. It seems very likely that even when the current crisis passes into history—almost certainly not until an effective vaccine or treatment becomes universally available—this trend could well turn out to be, to a significant extent, irreversible.

Retailers ranging in size from local grocery stores to national and regional chains have been moving aggressively to deploy robots capable of performing specialized tasks. For example, Brain Corporation, a maker of autonomous floor-scrubbing robots, has seen sales jump dramatically as the coronavirus crisis has put an urgent emphasis on the need for overnight deep cleaning in stores. Walmart expected to place the machines in over 1,800 of its U.S. stores by the end of 2020.[29] The retail giant also makes use of sorting machines that help organize newly arrived inventory by department as it is loaded off trucks. Likewise, retailers are investing in inventory-scanning robots that prowl the store aisles. Walmart planned to have the machines, which are six feet tall and equipped with fifteen cameras, autonomously inspect the shelves and scan product barcodes in at least a thousand of its stores by the summer of 2020.[30] The data collected by the robots is relayed to algorithms that track store inventory and immediately alert workers of the need to restock particular items. Analysis has shown that out-of-stock items directly correlate with lower in-store sales, so the inventory robots provide an immediate boost to profitability while offering customers a better experience. Indeed, machine learning algorithms are being used to manage everything from inventory levels to product selection to placement of particular items within stores.

All this allows physical retailers to begin taking advantage of the same kind of artificial intelligence that Amazon leverages so effectively in its online shopping business.

One of the hottest recent trends is the integration of so-called "mini–fulfillment centers" into the backs of traditional grocery stores. These facilities, which are set up by a number of startup companies, including Takeoff Technologies and Israeli-based Fabric, offer robotic fulfillment capabilities that are in many ways comparable to what is found in the much larger distribution centers built by companies like Ocado. The mini–fulfillment centers allow grocery stores to efficiently handle online delivery operations and can prepare up to 4,000 orders per week.[31] By keeping online operations separate from the main store, the technology allows grocers to avoid sending store employees into potentially crowded aisles to retrieve items while reducing pressure on inventory in customer areas of the store which may, in the age of the coronavirus toilet paper panic, already be running low. Though mini–fulfillment centers lack the economies of scale that give larger stand-alone warehouses a cost advantage, the upfront capital outlay and the time required to integrate them into existing stores is substantially reduced—important advantages for smaller chains or independent stores.

In general, the robots deployed in retail environments exhibit the same strengths and limitations as those found in warehouses or factories. The machines efficiently move and sort materials in the back of the store and navigate the aisles, scrubbing the floors or scanning product barcodes. What they can't do, for the time being, is actually stock the shelves. The fundamental limitation holding back a more widespread robotic revolution is more often than not dexterity. Just as robots can't yet pick a wide variety of items from shelves in a warehouse, they are yet not up to the even more exacting job of positioning products on retail shelves. This, of course, is destined to change as genuinely dexterous robots begin to arrive.

It's also important to note that the overall retail business model is changing. Most brick-and-mortar stores are under relentless pressure from Amazon and other online retailers, and it seems inevitable that sales will continue to gradually shift away from traditional retail environments and toward the massive, and ever more automated, distribution centers run by e-commerce providers. Even in the grocery sector, online ordering and delivery is growing in popularity and has been dramatically accelerated by the need for nearly everyone to stay home during the height of the coronavirus crisis. Time will tell whether this shift in consumer preferences proves permanent, but it seems likely that once customers become accustomed to the convenience of having groceries delivered to their doorstep, the transformation will be fairly durable. That could lead to a general restructuring of retail grocery stores—so that the automated operations in the back of the store become relatively more important and both the floor space and product inventory dedicated to customer shopping aisles gradually shrink. Eventually, we might see the emergence of grocery stores that are essentially warehouses offering nearly instant fulfillment for either delivery or pickup, perhaps with small areas where customers can view products on display before ordering through a kiosk or mobile device.

One especially important trend in retail automation doesn't require any robotic dexterity or indeed any moving parts at all. In an entirely new retail model—cashierless stores—shoppers simply walk in, grab items from the shelves and then leave without ever encountering a checkout line, cashier or even an explicit payment mechanism. The concept first emerged with Amazon's Go convenience stores in 2018. Customers enter the roughly 2,000-square-foot stores by first activating an app on their smartphone and then scanning it as they pass through a subway station–like turnstile. Once in the store, they simply remove items from the shelf and place them directly in their shopping bag. All this is enabled by a remarkable synthesis of sensors and cameras clustered on the ceiling throughout the store. While Amazon is secretive about the details, the cameras are able to accurately track products as they are taken from the shelves, and the data is processed by deep learning systems that use image recognition capability to reliably record the purchases of every customer in the store as he or she moves through the aisles selecting items.

The technology is not perfect, and some losses do occur, but it is remarkably difficult to intentionally deceive the system. Customers can, for example, take an item then put it back on the shelf, perhaps in a different spot, and then retrieve it again, and the purchase will still be correctly tabulated. Even overt attempts to shoplift, such as by obscuring an item before removing it or quickly placing an item in a pocket rather than the shopping bag rarely succeed. Once a shopper leaves the store, again exiting through a turnstile, the purchases are automatically charged to the customer's Amazon account.[32]

Amazon has established Go convenience stores in twenty-six locations in major U.S. cities, and by one report is considering eventually opening as many as 3,000 stores across the United States.[33] In February 2020, the company announced its first full-size cashierless grocery store. Located in the Capitol Hill suburb of Seattle, the supermarket is about 10,000 square feet and stocks roughly 5,000 items. While Amazon is, as usual, the highest-profile player, a number of startup companies are racing to bring similar technologies to market. Accel Robotics, for example, received $30 million in venture capital to fund its "grab and go" technology in December 2019. Other startups include Trigo, Standard Cognition and Grabango, all of which have raised at least $10 million from investors.[34] Amazon has reportedly also licensed its technology to other retailers.[35] In other words, we are on the verge of seeing a vibrant and highly competitive market for the technology that powers stores without checkout aisles. Given this, it's a good bet that a variety of existing retailers will move toward adopting the new model.

If cashierless stores do gain traction, they have the potential to unleash a major industry disruption and eventually put the jobs of more than three and a half million cashiers in the United States alone at significant risk. Beyond the increase in convenience and the time saved by not standing in checkout lines, these stores may be an especially good fit in a future shaped by the coronavirus because they offer completely touch-free payment without the need to ever come in close proximity to a human worker. Ironically, Amazon temporarily shut down most of its Go stores as the coronavirus unfolded, perhaps because the stores are so popular that they attract long lines of shoppers. However, in the long run, the technology seems ideally suited to a world where social distancing is, at least for a time, at a premium.

Another sector where I think robotic automation will have a significant impact in the relatively near future is the fast food industry. McDonald's, for example, has been undertaking a major push to install touchscreen ordering kiosks in its restaurants worldwide. The company reportedly spent nearly a billion dollars on the machines in 2019 and expected to install them in nearly all its U.S. locations.[36] The automated kiosks are already ubiquitous in McDonald's European restaurants.

The jobs in the back of the restaurant cooking and preparing food are also likely to see increased automation in the near future. These jobs have already been largely deskilled and divided into a series of highly routine tasks. This is part of an industry strategy to keep wages low and adapt to employee turnover rates that were as high as 150 percent in 2019.[37] The mechanized nature of these jobs makes it highly feasible to gradually substitute automated machines for workers.

One of the most successful examples so far is San Francisco–based Creator, Inc. The sophisticated and aesthetically designed robot at the company's first restaurant in the city's South of Market area is able to crank out a gourmet-quality hamburger every thirty seconds. Customers customize and order their burger using a mobile app. The robot then completely automates production of the hamburger from start to finish. No human being ever touches the food during the process. And the machine adds twists that you might not find even in high-end restaurants staffed by human cooks. The meat is freshly ground and the cheese freshly grated for each burger; buns are sliced and vegetables are cut to order. Creator sells its burgers for $6—about half of what you might expect to pay for similar quality at other restaurants. The company's strategy is not to build a cheap robotic hamburger, but rather to reduce labor costs as a means of investing more in food quality. Creator allocates about forty percent of its costs to food, while a typical restaurant might spend thirty percent.[38]

It turns out that developing and building a machine capable of fully automating the production of gourmet-quality hamburgers is not a trivial undertaking. Creator was founded in 2012, and I wrote about the company, then called Momentum Machines, in my 2015 book *Rise of the Robots*. It took more than six years of hardware and software engineering, design and testing before the robot was ready to be put into production, and the San Francisco location was opened in June 2018. However, the company, which has received funding from Google Ventures and other top Silicon Valley venture capital firms, may now be poised to rapidly expand or perhaps license its technology to other restaurants.

Creator, with its strategy of leveraging automation to produce high-end hamburgers, will likely soon be joined by a variety of other startup companies that are instead developing robots to produce cheap commoditized burgers. Eventually, I think, it's inevitable that the major fast food chains, as well as smaller independent restaurants, will begin to introduce these technologies. Once one major player does so and is able to capitalize on the technology, a competitive dynamic virtually guarantees widespread automation.

Nor will the impact be limited to hamburgers. Entrepreneurs will find effective ways to deploy robots in the production of everything from pizza to tacos to your favorite coffee drink. And, of course, the conventional wisdom that customers have a strong preference for interaction with human employees over robots in these kinds of environments could well be turned on its head to some extent in the wake of the coronavirus. Suddenly a machine that can produce fully prepared food with a total absence of human contact may offer significant marketing advantages. As I write this, restaurants across the world have been largely limited to takeout service. In the event that consumer preference undergoes a permanent shift in favor of takeout dining as the crisis continues to unfold, that would tend to further minimize any advantages offered by human interaction, alter the business models and cost structures of restaurants and quite possibly accelerate the transition to automation across the industry.

ARTIFICIAL INTELLIGENCE IN HEALTHCARE

Over the half-century from 1970 to 2019, healthcare spending as a percentage of gross domestic product in the United States more than doubled, from about seven percent to roughly eighteen percent.[39] The upward slope of the healthcare expenditure graph in other developed countries is not as extreme, and the current spending numbers are lower than in the U.S., but the story is broadly similar. In countries including Germany, Switzerland and the United Kingdom, for example, spending as a fraction of GDP has at least doubled over the same period.[40] The primary driver of this global trend is what's known as "cost disease" or the Baumol effect, a phenomenon researched by the economists William Baumol and William Bowen, who described it in a 1966 book that focused on cost disease in the performing arts sector.[41]

The main idea underlying cost disease is that certain sectors of economy, most notably healthcare and higher education, require non-routine, non-scalable efforts by highly skilled workers, and as a result these sectors have not seen the productivity increases that have manifested in the broader economy. As automation has advanced relentlessly in factories, for example, the efforts of an individual manufacturing worker have been vastly amplified. The same has been true in sectors like retail and fast food, where the introduction of new technology as well as more efficient workplace organization, management techniques and business models—including the advent of "big box" stores and online shopping—have likewise boosted productivity. In healthcare, however, patients continue to require highly individualized attention from doctors, nurses and other skilled professionals. To be sure, new knowledge and technology have increased the quality of care and produced vastly better patient outcomes, but so far, this has not amplified the efforts of healthcare workers in the way that we have seen with factory workers. Nonetheless, wages in the healthcare sector have had to rise to keep pace with what workers in more productive industries earn. Without this, doctors and nurses would be likely to leave (or never enter) their professions in favor of more attractive opportunities elsewhere. The result is that healthcare costs have come to dominate an ever larger share of the economy.[42]

One of the greatest opportunities—and challenges—for artificial intelligence is to find a cure for healthcare's cost disease. Will artificial intelligence prove to be the technology that will finally bend the healthcare spending curve by scaling productivity increases across the industry? It hasn't happened yet, but there are certainly good reasons to be optimistic that AI will have a significant impact over the long run.

Robots have already made significant inroads in hospitals, but they are subject to the same basic limitations that we've seen in warehouse and retail environments. Disinfecting robots, for example, are rapidly growing in popularity. These machines are able to create a virtual map of a room in a hospital and then autonomously navigate while directing intense ultraviolet radiation at every surface. Unlike a human worker, the robot never misses a spot. The UV light rapidly destroys the RNA or DNA in viruses and bacteria, allowing a typical room to be disinfected in fifteen minutes or so. The procedure has been shown to be significantly more effective then liquid disinfectants, especially since some of the most dangerous "superbugs" have evolved to be resistant to these chemicals. One manufacturer, San Antonio–based Xenex, saw a 400 percent increase in demand for its disinfecting robots in the first three months of the coronavirus pandemic.[43]

Other robots autonomously navigate the hallways and elevators in hospitals, delivering drugs, linens and medical supplies. The robots are able to carry heavy loads and periodically return to charging stations to have their batteries topped up. Likewise, massive pharmacy robots that prepare and dispense thousands of prescriptions with flawless accuracy have increased efficiency and reduced medication errors in major hospitals. The machines completely automate the process; from the time a physician enters the order in a hospital's computer system, no human touches the medication until after it is packaged and labeled with a tracking barcode by the robot. The system also keeps track of the pharmacy's inventory and automatically generates orders for new medications on a daily basis.[44]

These are important advances, but once again, they're limited to the most routine aspects of the work that needs to be performed in healthcare environments. There are no robots that can scale the highly skilled interventions required of doctors and nurses. Surgical robots like the da Vinci system have become very popular and may amplify the capabilities of surgeons, but these machines are not autonomous. Instead, the same doctor that would otherwise perform the surgery manually now manipulates the robot. The patient may be happier with the result, but the time required of the surgeon and the accompanying medical team are not dramatically reduced. The manipulative work performed by doctors and nurses presents an extraordinary challenge for artificial intelligence because it requires extreme dexterity combined with problem solving and interpersonal skills, as well as the ability to handle an unpredictable environment where every situation, and every patient, is unique. As far as physical healthcare robots are concerned, the productivity scaling effect that we have seen in factories or warehouses likely lies in the distant future and will require not just vastly improved robotic dexterity, but quite possibly artificial general intelligence or something very close to it.

Given the limitations of physical robots, it seems likely that any truly significant near-term AI impact on healthcare will emerge in activities that require no moving parts. In other words, artificial intelligence will make its mark in the processing of information and in purely intellectual endeavors, such as diagnosis or the development of treatment plans. The interpretation of medical images using machine vision techniques is an especially promising area. A number of studies have demonstrated that deep learning systems are, in many cases, able to match or exceed the capabilities of human radiologists. For example, a study published by a team of researchers from Google and several medical schools in 2019 showed that a deep learning system was able to beat radiologists at diagnosing lung cancer by analyzing CT scans. Google's system was 94.4 percent accurate and "outperformed all six radiologists" in cases in which a prior CT scan for the patient was not available, and "was on-par with the same radiologists" when a previous image was available for comparison.[45]

Likewise, radiology AI systems were employed on an emergency basis in some cases as the coronavirus pandemic threatened to overwhelm hospitals. Amidst a shortage of tests for COVID-19, chest X-rays that showed evidence of the pneumonia often caused by the virus became an important alternative diagnostic technique. Some hospitals experienced backlogs resulting in delays of six hours or more as radiologists struggled to analyze the images. In response, two manufacturers of AI diagnostic tools, Mumbai-based Qure.ai and the Korean company Lunit, were able to rapidly recalibrate their systems to focus on the coronavirus. One study found that Qure.ai's system was ninety-five percent accurate in distinguishing COVID-19 from other conditions that cause pneumonia.[46]

Results like these have led to enthusiasm that can sometimes blur into hype, and among some deep learning experts it's often taken almost as a given that AI systems will completely replace human radiologists in the relatively near future. Turing Award winner Geoffrey Hinton, arguably the most prominent advocate of deep learning, said in 2016 that "we should stop training radiologists now" because "it's just completely obvious that within five years deep learning is going to do better than radiologists." Hinton compared the doctors to Wile E. Coyote, the *Roadrunner* cartoon character famous for often finding himself "already over the edge of the cliff" before looking down and only then plunging into the abyss.[47] As I write this, four years after Hinton's statement, there's no evidence of looming unemployment for radiologists. Indeed, practitioners push back aggressively against the argument that their profession will soon evaporate. In September 2019, Alex Bratt, a doctor in the Department of Radiology at Stanford Medical School, published a commentary entitled "Why Radiologists Have Nothing to Fear from Deep Learning" in which he made the case that deep learning–powered radiology systems lack flexibility and holistic reasoning and are generally limited to simple cases. The systems, he wrote, have no ability to integrate information from "clinical notes, laboratory values, prior images" and the like. As a result, the technology has so far excelled only with "entities that can be detected with high specificity and sensitivity using only one image (or a few contiguous images) without access to clinical information or prior studies."[48] I suspect that Geoff Hinton would argue that these limitations will inevitably be overcome, and he will very likely turn out to be right in the long run, but I think it will be a gradual process rather than a sudden disruption.

An additional reality is that there are a variety of challenging hurdles beyond the capability of the technology itself that will probably make it very difficult to send radiologists—or any other medical specialists—to the unemployment line anytime soon. Nearly every aspect of healthcare is heavily regulated, sometimes by multiple entities with overlapping authority. Taking licensed physicians completely out of the loop is not going to be easily accomplished. The power of organizations like the American Medical Association gives doctors far more influence over their fate than most other types of workers. There are also important liability issues. An error that leads to a bad outcome for a patient can easily lead to a malpractice lawsuit. Currently this liability is distributed among thousands of individual doctors. If the work is instead performed by a device or algorithm developed and marketed by a deep-pocketed corporation, that would concentrate the liability and potentially create an incentive for a deluge of litigation. These are all issues that may be resolved in the long run, but for the foreseeable future, I think the question is not whether AI will replace radiologists but whether it can significantly boost their productivity. If deep learning allows radiologists to analyze significantly more images over a given time frame while offering an instant second opinion that minimizes the error rate, that will amplify the efforts of individual doctors and may, over time, lead to medical students choosing a different specialty in response to natural market demand for their services.

Visual images are, of course, not the only form of information accessible to deep learning algorithms. The transition to electronic medical records has generated a massive trove of data that is in many ways ideally suited to the application of artificial intelligence. Leveraging this resource in ways that improve efficiency, cut costs and result in better patient outcomes is probably the single most promising near-term opportunity for AI in healthcare. By some accounts, medical errors are the third leading cause of death in the United States, outstripped only by cancer and heart disease. As many as 440,000 Americans die each year as the result of preventable errors.[49] Mishaps resulting from the administration of an incorrect medication or the wrong dosage are especially prevalent.

In a 2019 study, an AI application from the Israeli startup MedAware was turned loose on a historical database of nearly 750,000 patient interactions that had occurred at Brigham and Women's Hospital in Boston during 2012 and 2013. The system flagged nearly 11,000 errors. Analysis of the results showed that MedAware's software was ninety-two percent accurate in uncovering legitimate errors, that nearly eighty percent of the alerts offered valuable clinical information and that more than two thirds of these mishaps would not have been identified with the existing systems in use at the hospital. In addition to improved patient outcomes and potential lives saved, the study found that Brigham and Women's would have saved about $1.3 million in treatment costs that resulted directly from the errors.[50]

One of the highest profile applications of artificial intelligence to patient data occurred in 2016 when DeepMind entered into a five-year data-sharing agreement with the U.K.'s National Health Service. The NHS provided DeepMind access to information on over a million patients. The pilot applications developed included a system that could analyze patient records and test results and then instantly alert NHS staff when a patient was in danger of an acute kidney injury, as well as an AI system that proved able to diagnose eye disease from medical scans with an accuracy that in some cases exceeded that of doctors. Though progress was promising, the arrangement exploded into controversy in 2019 when the program was transferred to DeepMind's parent company, Google. There was an immediate backlash against the specter of the tech giant having access to NHS patient data despite the fact that Google claimed strict privacy policies were in place and the data was carefully anonymized.[51] All this illustrates, once again, how factors beyond the capability of the technology itself—in this case, perceived privacy concerns—can act to significantly slow the deployment of artificial intelligence in the healthcare arena.

Some of the most surprising successes with artificial intelligence in healthcare are occurring in the mental health arena. Woebot Labs, a Silicon Valley startup founded in 2017, has developed a chatbot powered by natural language processing technology similar to what is used in Alexa and Siri, combined with carefully scripted conversational elements developed by psychologists. Woebot's approach is essentially to automate cognitive behavioral therapy, or CBT, a proven technique for helping

people with depression or anxiety. Within a week of the chatbot's release, more than 50,000 people conversed with the application. As founder and CEO Alison Darcy points out, "Woebot
can be there at 2 a.m. if you're having a panic attack and no
therapist can, or should be, in bed with you."[52] Indeed, the chatbot's unlimited twenty-four-hour availability, currently free of
charge, is something entirely new in mental health therapy, and
the application is already filling a critical space. Even workers
who have health insurance coverage in the United States often
have limited access to mental health services. The situation is far
worse in many developing countries with substandard healthcare systems. In regions where governments struggle to provide
the population with even basic medical care, getting access to a
mental health professional is a near impossibility for most citizens. Woebot regularly converses with people in more than 130
countries, many of whom communicate by using AI-powered
translation tools to interface with the chatbot's English-only
service.[53] In a world where a mental health crisis is becoming
increasingly evident, and has likely been greatly exacerbated by
the additional stress and anxiety brought on by the coronavirus
pandemic, tools like this offer what may be, for many people,
the only viable solution. I think it is somewhat ironic that the
specific field of healthcare that we might naturally regard as being the most intrinsically human is also the first area to benefit
from the kind of scalable AI-driven productivity improvements
that we someday hope will transform the industry as a whole.

The most important foreseeable and genuinely disruptive
breakthrough in medical artificial intelligence may turn out to
be the advent of a comprehensive and reliable system oriented
toward general diagnosis and treatment—in other words, a kind
of "doctor in a box." The point would not be to replace doctors
but rather to augment them in a way that effectively democ-
ratizes the skill and experience of the very best physicians. It's

easy to imagine a future where a powerful diagnostic AI system dramatically increases the productivity of doctors while creating an environment in which even an inexperienced or mediocre physician navigates patient encounters with what amounts to a virtual team of elite specialists looking over his or her shoulder providing continuous advice.

We are definitely not there yet, and one of the earliest attempts to move along this path offers a cautionary tale. Immediately after Watson's triumph in the 2011 *Jeopardy!* challenge, IBM moved aggressively to repurpose the technology for healthcare and other industries and built a new billion-dollar business unit around Watson. IBM's vision was that Watson would assimilate knowledge from a wide variety of sources; it would inhale a torrent of information from textbooks, clinical notes, diagnostic and genetic test results and scientific papers, and then leverage a superhuman ability to connect the dots in ways that would elude even the most capable expert. IBM hoped the technology would deliver tangible benefits in applications like the development of personalized treatment plans for complex diseases like cancer. Despite extreme hype and glowing media pieces declaring that Watson was "going to medical school" and preparing to "take on cancer"[54] as though it were the next *Jeopardy!* match, the results have, at least so far, been underwhelming. In 2017 the MD Anderson Cancer Center at the University of Texas, one of IBM's most highly touted healthcare partnerships, discontinued working with Watson after finding no real benefits from the technology.[55] Still, IBM remains confident and continues to invest in the idea, as do a growing number of other companies including both startups and giants like Google. The competition will continue to be keen as the return on an investment that leads to a truly successful technology is potentially staggering. I think it's inevitable that success will eventually arrive, but it likely will require AI technologies beyond current approaches in deep learning—or in other words, the kinds of breakthroughs in more general intelligence that researchers at the forefront of the field are pursuing. We'll cover the work going on at the AI frontier in Chapter 5.

Ultimately, if truly capable and robust systems do arrive, I think that could open the door to the emergence of a new class of medical professional. These might be people educated with perhaps an undergraduate or master's degree and trained specifically to interface between patients and an approved and regulated medical AI system. These lower-cost workers would not directly substitute for physicians but might work under their supervision and have the ability to take on more routine cases. Family doctors in the United States, for example, are generally inundated by a constant stream of patients with the same chronic conditions, most notably obesity, high blood pressure and diabetes. A new class of practitioner working hand in hand with artificial intelligence might go a long way to lessening this burden while also expanding geographic coverage. Many rural areas of the U.S. already have serious shortages of doctors, and this will only worsen as our population ages. To address these issues and eventually achieve the kind of productivity increases that will finally rein in healthcare's cost disease, I think we will have little choice except to rely far more heavily on medical machine intelligence.

SELF-DRIVING CARS AND TRUCKS: A LONGER THAN EXPECTED WAIT

Elon Musk's promise of a million robotic taxis operating on roads by the end of 2020 is only the most recent example of overexuberance in the autonomous vehicle industry. Perhaps because of the centrality of the automobile to our way of life, especially in the United States, no application of artificial intelligence has been subject to as much hype and hyperbolic enthusiasm as the self-driving car. Since the industry's emergence following the Defense Advanced Research Projects Agency (DARPA) grand challenges in 2004 and 2005, the technology has achieved astonishing progress while at the same time regularly falling short of overinflated expectations. In 2015, it was widely predicted by the most knowledgeable industry insiders that fully autonomous vehicles would be on our roads within five years. Chris Urmson, one of the pioneers of the field, who was formerly the chief technology officer for Google's self-driving car spinoff, Waymo, and is now CEO and founder of the autonomous driving startup Aurora, famously speculated that his then-eleven-year-old son might have no need to pursue a driver's license when he turned sixteen. Major manufacturers including Toyota and Nissan likewise promised self-driving vehicles by 2020.[56] All those predictions have now been rolled back. Urmson remains confident and said in 2019 that he expects at least "hundreds" of fully autonomous vehicles to be deployed on public roads within five years,[57] and that there may be 10,000 or more such cars operating within ten years.[58] My own view is that even those predictions could well turn out to be optimistic. I'd say there's a real danger that truly autonomous cars are going to remain five years in the future for many years to come.

The reality is that the routine operation of autonomous cars on both highways and in more urban environments—in other words, situations where things work more or less as expected—has largely been solved. If public roads were anything like the inside of an Amazon warehouse in terms of the overall level of predictability, self-driving cars might already be widely deployed.

The problem, of course, is in the so-called edge cases, the virtually infinite number of unusual interactions and situations that are difficult or impossible for a self-driving car to accurately predict or, in many cases, to correctly interpret. Most self-driving car initiatives depend on highly precise advanced mapping of the streets being traveled. Therefore, unexpected road closings, construction or traffic accidents can create problems. Inclement weather, especially heavy rain or snow, also produces major impediments. But the greatest challenge may be to safely interact with an ecosystem populated by unpredictable pedestrians, bicyclists and drivers. In cities like San Francisco, it's not uncommon to encounter pedestrians who are distracted or drunk. Even those who are alert often act in ways that are a challenge to interpret, stepping tentatively off the curb in some cases, or in certain neighborhoods, and far more aggressively in others. In densely populated areas, much of the coordination between drivers and pedestrians relies on social interactions that would be very difficult for a self-driving car to understand or replicate. A connection achieved through eye contact, a wave of a hand, pausing midstride to wait for a driver's acknowledgement and numerous other nuanced behaviors make up a kind of unspoken language that is somehow understood by nearly everyone who shares the road. I think it is quite possible that it may turn out that negotiating these types of interactions is simply beyond the capabilities of today's deep learning systems. In other words, truly autonomous cars may require technology much further along the path toward general machine intelligence, and that could be a long wait.

Many analysts believe that, given the difficulties faced by autonomous cars in urban settings, the first truly practical driverless vehicles to appear on our roads will be long-haul trucks. Driving on highways, after all, is a problem that has largely already been solved by systems like Tesla's autopilot. While it's certainly true that the likelihood of an unpredictable event is lower on a highway than at a busy urban intersection, the consequences of an error are vastly magnified by the speeds involved and the fact that the vehicle is a fully loaded truck traveling with nearly unfathomable kinetic energy. And, of course, in spite of Elon Musk's exuberance, Tesla's autopilot system is in no way certified to operate without an attentive driver at the wheel. For these reasons I think it will be quite a while before we routinely see genuinely unmanned trucks on public highways.

I suspect the challenges faced by one small company may contain some important insights for the sector as a whole. In early 2017 I was invited to visit a San Francisco–based startup company called Starsky Robotics. The company's vision, as explained to me by its CEO and co-founder, Stefan Seltz-Axmacher, was to build a system capable of driving autonomously on highways over long distances but to have the trucks supervised by human operators via remote control. As the vehicles left or approached the endpoints on their route or otherwise encountered more complex situations, the remote operator—generally a retrained truck driver—would drive the truck via a cellular connection from a video game–like console at the company headquarters. Seltz-Axmacher told me he believed the company would have fully autonomous, unmanned trucks on American roads within the next few years. Though I was greatly impressed by Starsky's team and the technology they showed me, I was very skeptical that they would achieve this, especially given the regulatory hurdles they would need to surmount. Nonetheless, Seltz-Axmacher and his team exceeded my expectations: the company successfully operated a driverless truck on a closed road in 2018, and then in 2019 became the first autonomous vehicle company to test a fully automated truck, with no safety driver on board, on a public highway.

Starsky also adopted a very innovative business model. Rather than competing directly with the growing number of well-funded startups hoping to develop and license the technology to enable autonomous driving, Starsky instead decided to directly enter the trucking business and use its system to gain a competitive advantage. The company's management believed that only by fully integrating development of the technology into the daily operations of a trucking company, and taking advantage of the flexibility to deploy the evolving system only in situations where it made sense, could near-term success be achieved.

Sadly, investors didn't ultimately buy into this vision, and the company was forced to shut down in early 2020 after failing to raise the next required round of venture capital. In a series of blog posts written after the company's demise, Seltz-Axmacher pointed to the limitations of deep learning as one of the primary challenges holding back progress in the industry. "Supervised machine learning doesn't live up to the hype," he wrote, "it isn't actual artificial intelligence" but rather "a sophisticated pattern-matching tool."[59] In other words, a system with the flexibility to offer truly autonomous driving under all circumstances, without the need for remote human supervision, may well be beyond the capability of today's deep learning systems and is unlikely to arrive in the near future. Seltz-Axmacher believes that the challenges faced by the industry are not yet fully appreciated and that investors missed an opportunity to safely put a fleet of self-driving trucks on highways in the near term, in part because of an overriding focus on the promise of full automation and on the more advanced features that were often demonstrated by competing startups, but were nowhere close to being ready for real-world deployment.

Developing sufficiently capable technology looms as the greatest challenge for the autonomous vehicle industry, but I think there are also some real questions regarding the potential business models for such vehicles. The natural place to deploy self-driving cars is generally assumed to be in ride-sharing services. Uber and its competitors have been subsidizing the cost of every ride by drawing on capital obtained through venture funding or, more recently, IPOs.[60] Given that this is unsustainable, self-driving cars are widely viewed as the long-term solution. Once the driver, who generally gets seventy to eighty percent of the fare, is out of the picture, the companies ought to have a smooth path to profitability. This is the primary reason that Uber views autonomous vehicle companies, particularly Waymo, as existential threats and chose to invest heavily in a self-driving program of its own beginning in 2016.

The problem with the assumption that self-driving technology will ride to their rescue is that Uber and Lyft are viewed as attractive internet-based businesses—and valued accordingly—because they act primarily as digital intermediaries, harvesting a slice of every transaction in return for providing software that automatically matches riders with drivers. This allows the companies to completely avoid the risky and unpleasant parts of the taxi business: stuff like owning, financing, maintaining and insuring vehicles. All of that gets pushed onto the drivers. No oil changes, car washes or flat tires for Uber or Lyft; they largely remain above the fray, hoovering up clean internet fees. Getting rid of the drivers, however, also means getting rid of the people who, rather conveniently, own and maintain the cars. Once the cars become autonomous, the companies will find themselves in the business of owning vast fleets of vehicles and will therefore be responsible for all the hassle and expense that comes along with that. Uber in effect will look quite a lot like Hertz or Avis—neither of which is valued as a "tech company." Moreover, the vehicles owned by the ridesharing companies will be far more expensive given the specialized equipment, such as lidar systems, that they require. In the aftermath of the coronavirus pandemic, there may also be far more emphasis placed on properly cleaning and sanitizing vehicles on a frequent basis. This, again, is something that is currently the responsibility of drivers.

I think it will be fascinating to watch the evolution of self-driving cars over the coming years in terms of both the technology and the business models that ultimately emerge. There are a large number of Silicon Valley startups focused on developing and licensing self-driving technology, as well as varying degrees of investment by virtually every major automotive manufacturer. A disruptive breakthrough could emerge from any of these initiatives, but I think one of the most interesting narratives will center on the widening gap between the strategies pursued by Waymo and Tesla, and how competition between these two companies plays out over time.

Waymo, the direct descendant of Google's self-driving car program initiated in 2009, has more experience than anyone else and is generally regarded as the industry leader. Waymo is the only company that offers an operating automated car service with which paying customers can already ride in a car with no driver at the wheel. This service, called Waymo One, is currently available only for predefined routes in a carefully mapped— or "geo-fenced"—region of suburban Phoenix. The roads are wide, the weather is predictably cooperative and pedestrians are sparse. In other words, the service is a far cry from hailing an Uber and going wherever you like in San Francisco or Manhattan. Nonetheless, Waymo One is an impressive feat, and I think it is more or less what self-driving car service will look like for the foreseeable future: specified routes with designated stops in carefully curated areas that aren't too challenging. Of course, this once again raises the question of how such a limited operation can become profitable. How inexpensive does a fully automated ride (in a very expensive vehicle) have to be to get a customer to opt for it over the far more flexible door-to-door service offered by a human-driven Uber or Lyft?

While Waymo proceeds deliberately and with laudable caution, Tesla, in contrast, continuously pushes the envelope, often transgressing into territory that many in the industry feel borders on reckless. The company has told its existing customers that their cars have all the necessary hardware to support fully autonomous driving, and that the capability will eventually be enabled through a software update. This is an extraordinarily ambitious promise. Tesla has also diverged from Waymo and virtually everyone else in the industry by forgoing lidar— systems that track objects around the car by firing a laser and then detecting the reflected light. Lidar is expensive and, at least in its current instantiation, ugly. Tesla uniquely believes that it can achieve full automation by relying solely on cameras and radar. As I noted previously, Tesla enjoys a significant advantage in terms of the data the multiple cameras on its cars collect. Waymo has a fleet of about 600 self-driving vehicles. Tesla has an expanding fleet of over 400,000 cars on the road collecting data. Waymo's vehicles have driven millions of miles on actual roads and billions of miles in simulation.[61] Tesla's cars have driven billions of real-world miles while operating under the control of its autopilot system. All this data collected on actual roads is a clear advantage, but ultimately success will depend on artificial intelligence that is sufficiently powerful to leverage that resource, and I think there are real questions as to whether today's deep learning technology is up to that task.

Another important question for the industry surrounds the level of autonomy that will ultimately be provided. Autonomous driving systems are divided into five categories. Levels 1 through 3 designate systems that are assistive in nature. The car can drive itself under limited circumstances, for example while cruising on a highway, but the driver must remain alert and ready to take control of the car at a moment's notice. Most automotive manufacturers, including Tesla, are focused on providing capability in this range. The problem is that because the system will work correctly virtually all the time, drivers will inevitably be lured into inattention. A number of Tesla drivers have told me, for example, that they routinely answer email on their phones while using their car's autopilot system in the car-pool lanes of Silicon Valley's freeways. This kind of behavior has already led to fatal accidents. It's unclear how the car can successfully enforce attentiveness on the part of the driver over long stretches of routine driving. One of the strongest selling points for self-driving systems is the promise that they will one day dramatically reduce the huge number of people—more than 1.3 million globally—who die each year in traffic accidents.[62] If systems that are merely assistive continue to come with dangers of their own, they may not be sufficient to put a meaningful dent in this number.

For this reason, Waymo, along with many of the other smaller startups in this space, have made a decision to focus ex-clusively on level 4 and 5 autonomy. This indicates a self-driving car in which you can go to sleep. Indeed, it may not have a brake pedal or a steering wheel. Here again, Tesla is a dramatic outlier. The company's claim that it can bridge the gap between these two visions, with a software update that will instantly upgrade its cars from level 2 to level 4 autonomy is, to say the least, remarkable. Many might say it's a promise that is completely over the top and little more than vaporware. I will be astonished if Tesla can achieve this anytime soon, but if the company can do it at all, it will, I think, position itself as the clear industry leader. Indeed, that expectation may to some extent already be factored into the price of the company's stock.

Elon Musk and the rest of Tesla's management team have clearly given a lot of thought to the prospects for full autonomy. Aside from the technology, they've also developed a potential solution to the business model problem. At the 2019 Autonomy Day event, Musk described a scheme in which Tesla owners would be able to have their cars participate in a robotaxi service run by the company. Tesla would get a cut of the ride-sharing fee in the same way that Apple generates revenue from its App Store. One interesting thing about this proposal is that it solves the ownership and maintenance problem that might eventually plague companies like Uber and Lyft. Tesla may have found a way to step into the role of a pure internet intermediary, while avoiding the need to own a fleet of cars. Most Tesla owners might not want to share their vehicles with strangers, but if the plan proves viable, many customers would presumably buy Tesla's vehicles as a business investment, rather than as personal cars.

There is little doubt that self-driving vehicles will someday be one of the most tangible and consequential manifestations of the revolution in artificial intelligence. The technology has the potential to reshape both our cities and our way of life while saving many thousands of lives. However, I think we will need to wait a decade or more before the technology really arrives. Strong evidence of the AI revolution will first emerge in other areas—places like warehouses, offices and retail stores—where the technical challenges are more manageable, the environment is more controllable, the technology is less subject to government regulation and the consequences of an error are far less dire. It is very exciting to imagine, however, that a single software update from Tesla could prove me wrong.

BLASTING OFF THE INNOVATION PLATEAU: SCIENTIFIC AND MEDICAL RESEARCH

Among those who might be described as "technoptimists," it is taken as a given that we live in an age of startling technological acceleration. The pace of innovation, we are told, is unprecedented and exponential. The most enthusiastic accelerationists—often acolytes of Ray Kurzweil, who codified the idea in his "Law of Accelerating Returns"—are confident that in the next hundred years, we will experience, by historical standards, the equivalent of something "more like 20,000 years of progress."[63]

Closer scrutiny, however, reveals that while the acceleration has been real, this extraordinary progress has been confined almost exclusively to the information and communications technology arena. The exponential narrative has really been the story of Moore's Law and the ever more capable software it makes possible. Outside this sector, in the world composed of atoms rather than bits, the story over the past half-century or so has been starkly different. The pace of innovation in areas like transportation, energy, housing, physical public infrastructure and agriculture not only falls far short of exponential, it might be better described as stagnant.

If you want to imagine a life defined by relentless innovation, think of someone born in the late 1800s who then lived through the 1950s or 1960s. Such a person would have seen systemic transformations across society on an almost unimaginable scale: infrastructure to deliver clean water and manage sewage in cities; the automobile, the airplane, jet propulsion and then the advent of the space age; electrification and the lighting, radios, televisions, and home appliances it later made possible; antibiotics and mass-produced vaccines; an increase in life expectancy in the United States from less than 50 years to nearly 70. A person born in the 1960s, in contrast, will have witnessed the rise of the personal computer and later the internet, but nearly all the other innovations that had been so utterly transformative in previous decades would have seen at best incremental progress. The difference between the car you drive today and the car that was available in 1950 simply does not compare to the difference between that 1950 automobile and the transportation options in 1890. And the same is true of a myriad of other technologies distributed across virtually every aspect of modern life.

The fact that all the remarkable progress in computing and the internet does not, by itself, measure up to the expectation that the kind of broad-based progress seen in earlier decades would continue unabated is captured in Peter Thiel's famous quip that "we were promised flying cars and instead we got 140 characters." The argument that we have been living in an age of relative stagnation—even as information technology has continued to accelerate—has been articulated at length by the economists Tyler Cowen, who published his book *The Great Stagnation* in 2011,[64] and Robert Gordon, who sketches out a very pessimistic future for the United States in his 2016 book *The Rise and Fall of American Growth*.[65] A key argument in both books is that the low-hanging fruit of technological innovation had been largely harvested by roughly the 1970s. The result is that we are now in a technological lull defined by a struggle to reach the higher branches of the innovation tree. Cowen is optimistic that we will eventually break free of our technological plateau. Gordon is much less so, suggesting that even the upper branches of the tree are perhaps denuded and that our greatest inventions may be behind us.

While I think Gordon is far too pessimistic, there is plenty of evidence to suggest that a broad-based stagnation in the generation of new ideas is quite real. An academic paper published in April 2020 by a team of economists from Stanford and MIT found that, across a variety of industries, research productivity has sharply declined. Their analysis found that the efficiency with which American researchers generate innovations "falls by half every 13 years," or in other words "just to sustain constant growth in GDP per person, the United States must double the amount of research effort every 13 years to offset the increased difficulty of finding new ideas."[66] "Everywhere we look," wrote the economists, "we find that ideas, and the exponential growth they imply, are getting harder to find."[67] Notably this extends even to the one area that has continued to generate consistent exponential progress. The researchers found that the "number of researchers required today to achieve the famous doubling of computer chip density" implied by Moore's Law "is more than 18 times larger than the number required in the early 1970s."[68] One likely explanation for this is that before you can push through the research frontier, you first have to understand the state of the art. In virtually every scientific field, that requires the assimilation of vastly more knowledge than has been the case previously. The result is that innovation now demands ever larger teams made up of researchers with highly specialized backgrounds, and coordinating their efforts is inherently more difficult than would be the case with a smaller group.

To be sure, there are many other important factors that might be contributing to the slowdown in innovation. The laws of physics dictate that accessible innovations are not distributed homogeneously across fields. There is, of course, no Moore's Law for aerospace engineering. In many areas, reaching the next cluster of innovation fruit may require a giant leap. Over- or ineffective government regulation certainly also plays a role, as does the short-termism that now prevails in the corporate world. Long-term investments in R&D are often not compatible with an obsessive focus on quarterly earnings reports or the coupling of short-term stock performance and executive compensation. Still, to the extent that the need to navigate increased complexity and an explosion of knowledge is holding back the pace of innovation, artificial intelligence may well prove to be the most powerful tool we can leverage to escape our technological plateau. This, I think, is the single most important opportunity for AI as it continues to evolve into a ubiquitous utility. In the long run, in terms of our sustained prosperity and our ability to address both the known and unexpected challenges that lie before us, nothing is more vital than amplifying our collective ability to innovate and conceive new ideas.

The most promising near-term application of artificial intelligence, and especially deep learning, in scientific research may be in the discovery of new chemical compounds. Just as DeepMind's AlphaGo system confronts a virtually infinite game space—where the number of possible configurations of the Go board exceeds the number of atoms in the universe—"chemical space," which encompasses every conceivable molecular arrangement, is likewise, for practical purposes, infinite. Seeking useful molecules within this space requires a multi-dimensional search of staggering complexity. Factors that need to be considered include the three-dimensional size and shape of the molecular structure as well as numerous other relevant parameters like polarity, solubility and toxicity.[69] For a chemist or materials scientist, sifting through the alternatives is a labor-intensive process of experimental trial and error. Finding a truly useful new chemical can easily consume much of a career. The lithium-ion batteries that are ubiquitous in our devices and electric cars today, for example, emerged from research that was initiated in the 1970s but produced a technology that could begin to be commercialized only in the 1990s. Artificial intelligence offers the promise of a vastly accelerated process. The search for new molecules is, in many ways, ideally suited to deep learning; algorithms can be trained on the characteristics of molecules known to be useful, or in some cases on the rules that govern molecular configuration and interaction.[70]

At first blush, this may seem like a relatively narrow application. However, the quest to find useful new chemical substances touches virtually every sphere of innovation. Accelerating this process promises innovative high-tensile materials for use in machines and infrastructure, reactive substances to be deployed in better batteries and photoelectric cells, filters or absorbent materials that might reduce pollution and a range of new drugs with the potential to revolutionize medicine.

Both university research labs and an expanding number of startup companies have turned to machine learning technology with enthusiasm and are already using powerful AI-based approaches to generate important breakthroughs. In October 2019, scientists at Delft University of Technology in the Netherlands announced that they were able to design a completely new material by exclusively relying on a machine learning algorithm, without any need for actual laboratory experiments. The new substance is strong and durable but also super-compressible if a force beyond a certain threshold is exerted on it. This implies that the material can effectively be squeezed into a small fraction of its original volume. According to Miguel Bessa, one of the lead researchers on the project, futuristic materials with these properties might someday mean that "everyday objects such as bicycles, dinner tables and umbrellas could be folded into your pocket."[71]

Such initiatives typically require researchers to have a strong technical background in artificial intelligence, but teams at other universities are developing more accessible AI-based tools that are poised to jump-start the discovery of new chemical compounds. Researchers at Cornell University, for example, are working on a project called SARA—Scientific Autonomous Reasoning Agent—which the team hopes will "dramatically accelerate, by orders of magnitude, the discovery and development of new materials,"[72] while researchers at Texas A&M are likewise developing a software platform designed to autonomously search for previously unknown substances.[73] Both projects are funded in part by the U.S. Department of Defense, an especially eager customer for any innovations that emerge. Just as cloud-based deep learning tools offered by Amazon and Google are democratizing the deployment of machine learning in many business applications, tools like these are poised to do the same for many areas of specialized scientific research. This will make it possible for scientists with training in areas like chemistry or materials science to deploy the power of AI without the need to first become machine learning experts. Artificial intelligence, in other words, is evolving into an accessible utility that can be wielded in ever more creative and targeted ways.

An even more ambitious approach involves integrating AI-based software geared toward the discovery of chemicals with robots that can perform physical laboratory experiments. One small company pushing in this direction is Cambridge, Massachusetts–based Kebotix, a startup that spun out of a leading materials science laboratory at Harvard, which has developed what it calls the "world's first self-driving lab for materials discovery." The company's robots can perform experiments autonomously, manipulating laboratory equipment like pipettes to transfer and combine liquids and accessing machines that perform chemical analysis. Experimental results are then analyzed by artificial intelligence algorithms, which in turn make predictions about the best course of action and then initiate more experiments. The result is an iterative, self-improving process that the company claims dramatically accelerates the discovery of useful new molecules.[74]

Many of the most exciting and heavily funded opportunities in the space where chemistry intersects with artificial intelligence are in the discovery and development of new drugs. By one account, as of April 2020, there were at least 230 startup companies focused on using AI to find new pharmaceuticals.[75] Daphne Koller, a professor at Stanford and the co-founder of the online education company Coursera, is one of the world's top experts on applying machine learning to biology and biochemistry. Koller is also the founder and CEO of insitro, a Silicon Valley startup, founded in 2018, that has raised over $100 million to pursue new medicines using machine learning. The broad-based slowdown in technological innovation that plagues the American economy as a whole is especially evident in the pharmaceutical industry. Koller told me that:

> The problem is that it is becoming consistently more challenging to develop new drugs: clinical trial success rates are around the mid-single-digit range; the pre-tax R&D cost to develop a new drug (once failures are incorporated) is estimated to be greater than $2.5 [billion]. The rate of return on drug development investment has been decreasing linearly year by year, and some analyses estimate that it will hit zero before 2020. One explanation for this is that drug development is now intrinsically harder: Many (perhaps most) of the "low-hanging fruit"—in other words, druggable targets that have a significant effect on a large population—have been discovered. If so, then the next phase of drug development will need to focus on drugs that are more specialized—whose effects may be context-specific, and which apply only to a subset of patients.[76]

The vision for insitro and its competitors is to use artificial intelligence to rapidly isolate promising drug candidates and dramatically cut development costs. Drug discovery, Koller says, is "a long journey where you have multiple forks in the road" and "ninety-nine percent of the paths are going to get you to a dead end." If artificial intelligence can provide "a somewhat accurate compass, think about what that would do to the probability of success of the process."[77]

Approaches like this are already paying dividends. In February 2020, researchers at MIT announced that they had discovered a powerful new antibiotic using deep learning. The AI system built by the researchers was able to sift through more than one hundred million prospective chemical compounds within days. The new antibiotic—which the scientists named halicin after HAL, the artificial intelligence system from *2001: A Space Odyssey*—proved lethal to nearly every type of bacteria it was tested against, including strains that are resistant to existing drugs.[78] This is critical because the medical community has been warning of a looming crisis of drug-resistant bacteria—such as the "superbugs" that already plague many hospitals—as the organisms adapt to existing medications. Because development costs are high and profits relatively low, few new antibiotics are in the development pipeline. Even those new drugs that have made it through the rigorous and expensive testing and regulatory approval process tend to be variations on existing antibiotics. Halicin, in contrast, seems to attack bacteria in a completely novel way, and experiments suggest that the mechanism may be especially resilient to the mutations that generally make antibiotics less effective over time. In other words, artificial intelligence has produced a solution based on the kind of "outside the box" exploration that is critical to meaningful innovation.

Another important milestone, also announced in early 2020, came from the U.K.-based startup company Exscientia, which used machine learning to discover a new drug for treating obsessive compulsive disorder. The company says the project's initial development took just one year—about one fifth the time that would be typical for traditional techniques—and claims it is the first AI-discovered drug to enter clinical trials.[79]

As we saw in Chapter 1, an especially notable achievement in the application of artificial intelligence to biochemical research was DeepMind's protein folding breakthrough announced in November 2020. Rather than attempting to discover a specific drug, DeepMind has instead deployed its technology to gain understanding at a more fundamental level. In late 2018, Deep-Mind entered an earlier version of its AlphaFold system in a biennial global contest known as the Critical Assessment of Structure Prediction, or CASP. Teams from around the world used a variety of techniques based on both computation and human intuition to attempt to predict the way proteins fold. AlphaFold won the 2018 contest by a wide margin, but even while prevailing, it was able to make the best prediction for only twenty-five of the forty-three protein sequences correctly. In other words, this preliminary version of AlphaFold was not yet accurate enough to be a truly useful research tool.[80] The fact that DeepMind was able to refine its technology to the point where a number of scientists declared the protein folding problem to be "solved" just two years later is, I think, an especially vivid indication of just how rapidly specific applications of artificial intelligence are likely to continue advancing.

Aside from using machine learning to discover new drugs and other chemical compounds, the most promising general application of artificial intelligence to scientific research may be in the assimilation and understanding of the continuously exploding volume of published research. In 2018 alone, more than three million scientific papers were published in more than 40,000 separate journals.[81] Making sense of information on that scale is so far beyond the capability of any individual human mind that artificial intelligence is arguably the only tool at our disposal that could lead to some sort of holistic comprehension.

Natural language processing systems based on the latest advances in deep learning are being deployed to extract information, identify non-obvious patterns across research studies and generally make conceptual connections that might otherwise remain obscure. IBM's Watson technology continues to be one important player in this space. Another project, Semantic Scholar, was initiated by the Seattle-based Allen Institute for Artificial Intelligence in 2015. Semantic Scholar offers AI-enabled search and information extraction across more than 186 million published research papers in virtually every scientific field of study.[82]

In March 2020, the Allen Institute joined with a consortium of other organizations including Microsoft, the National Library of Medicine, the White House Office of Science and Technology, Amazon's AWS division and others to create the COVID-19 Open Research Dataset, a searchable database of scientific papers relating to the coronavirus pandemic.[83] The technology enables scientists and healthcare providers to rapidly access answers to specific questions in a broad range of scientific areas, including the biochemistry of the virus, epidemiological models, and treatment of the disease. As of April 2021, the database contained more than 280,000 scientific papers and was being heavily used by scientists and doctors.[84]

Initiatives like these have enormous potential to be crucial tools in accelerating the generation of new ideas. The technology remains in its infancy, however, and real progress will likely require surmounting at least some of the hurdles on the path to more general machine intelligence, a subject we'll delve into in Chapter 5. It's easy to imagine that a truly powerful system could step into the role of an intelligent research assistant for scientists, offering the ability to engage in genuine conversation, play with ideas and actively suggest new avenues for exploration.

Still, I think it's important to maintain a measured and realistic view of what might eventually be possible. None of this implies that artificial intelligence will be a panacea for turbocharging innovation or that we should expect results to consistently be achieved on an accelerated time frame. Science is, after all, fundamentally about experimentation, and conducting and evaluating the outcomes of experiments takes time. In some cases, the scientific method can indeed be accelerated, perhaps through the use of laboratory robots or even by performing some experiments at high speed in simulated environments.

In fields like medicine and biology, however, experiments often must be conducted within living organisms, and here, the potential for dramatically speeding up the process is quite limited. The successful quest for COVID-19 vaccines brings this reality into sharp focus. Scientists were able to formulate vaccine candidates within weeks of obtaining the virus's genetic code. The long wait for serviceable vaccines was almost entirely due to the need for extensive testing in both animals and humans, along with the need to ramp up manufacturing capacity to produce the billions of doses that are required. The truth is that even if we had access to truly advanced, science fiction–level artificial intelligence, it is not at all clear that the technology could have delivered a vaccine in a dramatically shorter time frame. This is one of the reasons that I'm skeptical of Kurzweillian claims that artificial intelligence will soon lead to a dramatic lengthening of the human lifespan. Even if AI does help generate powerful new ideas in this space, how will we test any resulting treatments for both safety and efficacy without waiting many years or even decades for conclusive results? To be sure, there are many opportunities for regulatory reform that might streamline the approval of new drugs and treatments, but at the end of the day, even the most intelligent and creative scientists must wait for the experimental results that confirm the veracity of their ideas.

THE INTENT OF this chapter has been to offer a brief tour of some of the most interesting and consequential applications of artificial intelligence while highlighting areas where AI seems likely to be disruptive in the near term and where we may have a longer wait. The list is nowhere close to being exhaustive.

Artificial intelligence will eventually touch and transform virtually everything.

The argument that artificial intelligence is rapidly evolving into an electricity-like utility effectively captures the potential reach and transformative nature of the technology. Compared to electricity, however, AI is a vastly more complex and dynamic technology that will continuously improve while delivering a nearly limitless number of ever-changing capabilities. In order to understand the true potential of this new utility, we need to delve into the science and history of artificial intelligence and see how the field is evolving, the challenges that lie ahead and the competing ideas that will shape the technology as it continues to progress. All this will be the subject of the next two chapters.

CHAPTER 4

THE QUEST TO BUILD INTELLIGENT MACHINES

THE A.M. TURING AWARD IS GENERALLY RECOGNIZED AS THE "Nobel Prize" of computing. Named after the legendary mathematician and computer scientist Alan Turing and awarded annually by the Association for Computing Machinery, the Turing Award represents the pinnacle of achievement for those who have devoted their careers to advancing the state of the field. Like the Nobel, the Turing prize comes with a $1 million financial award, which is funded primarily by Google.

In June 2019, the 2018 Turing Award was awarded to three men—Geoffrey Hinton, Yann LeCun and Yoshua Bengio—in recognition of their lifetime contributions to the advancement of deep neural networks. This technology—also known as deep learning—has, over the past decade, revolutionized the field of artificial intelligence and produced advances that just a short time ago would have been considered science fiction.

Tesla drivers routinely let their cars navigate highways autonomously. Google Translate instantly produces usable text, even in obscure languages that few of us have heard of, and companies like Microsoft have demonstrated real-time machine translation that renders spoken Chinese into English. Children are growing up in a world where it is routine to converse with Amazon's Alexa, and parents are worrying about whether these interactions are healthy. All of these advances—and a multitude of others—are powered by deep neural networks.

The basic idea underlying deep learning has been around for decades. In the late 1950s, Frank Rosenblatt, a psychologist at Cornell University, conceived the "perceptron," an electronic device that operated on principles similar to those of biological neurons in the brain. Rosenblatt showed that simple networks made up of perceptrons could be trained to perform basic pattern recognition tasks, such as deciphering images of numerical digits.

Rosenblatt's initial work on neural networks generated enthusiasm, but as significant progress failed to materialize, the technique was eventually pushed aside by other approaches. Only a small group of researchers, including especially the three winners of the 2018 Turing Award, continued to focus on neural networks. Among computer scientists, the technology came to be viewed as a research backwater and a likely career dead end.

Everything changed in 2012 when a team from Geoff Hinton's research lab at the University of Toronto entered the ImageNet Large Scale Visual Recognition Challenge. In this annual event, teams from many of the world's leading universities and corporations competed to design an algorithm that could correctly label images selected from a massive database of photographs. While other teams used traditional computer programming techniques, Hinton's team unleashed a "deep" (or many-layered) neural network that had been trained on thousands of example images. The University of Toronto team blew the doors off the competition, and the world woke up to the power of deep learning.

In the years since, virtually every major technology company has made massive investments in deep learning. Google,

Facebook, Amazon and Microsoft, as well as the Chinese tech leaders Baidu, Tencent and Alibaba, have made deep neural networks absolutely central to their products, operations and business models. The computer hardware industry is also being transformed, with companies like NVIDIA and Intel competing to develop computer chips that optimize the performance of neural networks. Experts in deep learning command seven-figure compensation packages and are treated like star athletes as companies compete for a limited pool of talent.

Though advances in artificial intelligence over the past decade have been both extraordinary and unprecedented, this progress has largely been driven by scaling up to ever larger troves of data being gobbled up by neural learning algorithms running on faster and faster computer hardware. There's a growing sense among AI experts that this approach is not sustainable and that entirely new ideas will need to be injected into the technology in order to continue carrying things forward. Before delving into the possible future of AI, let's briefly look at how it all began, the path that the field has traced up until now and how the deep learning systems that have produced such revolutionary progress over the past few years actually work. As we'll see, since its earliest days, research into artificial intelligence has been marked by competition between two entirely different approaches to building smart machines. Tension between these two schools of thought is once again coming to the forefront and may well be poised to shape the way the field progresses in the coming years and decades.

CAN MACHINES THINK?

Machines with the ability to think and act like humans inhabited the imagination long before the invention of the first electronic computers. In 1863, the English author Samuel Butler wrote a letter to the editor of the local newspaper in Christchurch, New Zealand. Entitled "Darwin Among the Machines," the letter envisioned "living machinery" that might someday evolve to match and perhaps even supplant human beings. Butler called for an immediate war against this emerging mechanical species, proclaiming that "every machine of every sort should be destroyed,"[1] a concern that seems a bit premature given the state of information technology in 1863, but which sketched out a narrative that has been repeated again and again, most recently in movies like *The Terminator* and *The Matrix*. Nor are Butler's fears confined to science fiction. Recent advances in AI have led prominent figures like Elon Musk and the late Stephen Hawking to warn of scenarios remarkably similar to what Butler worried about more than 150 years ago.

Opinions differ as to exactly when artificial intelligence became a serious field of study. I would mark the origin as 1950. In that year, the brilliant mathematician Alan Turing published a scientific paper entitled "Computing Machinery and Intelligence" that asked the question "Can machines think?"[2] In his paper Turing invented a test, based on a game that was popular at parties, which is still the most commonly cited method for determining if a machine can be considered to be genuinely intelligent. Turing, born in London in 1912, did groundbreaking work on the theory of computation and the nature of algorithms, and is generally regarded as the founding father of computer science. Turing's most important accomplishment came in 1936, just two years after he graduated from the University of Cambridge, when he laid out the mathematical principles for what is today called a "universal Turing machine"—essentially the conceptual blueprint for every real-world computer that has ever been built. Turing clearly understood at the very inception of the computer age that machine intelligence was a logical and perhaps inevitable extension of electronic computation.

The phrase "artificial intelligence" was coined by John McCarthy, who was then a young mathematics professor at Dartmouth College. In the summer of 1956, McCarthy helped arrange the Dartmouth Summer Research Project on Artificial Intelligence at the college's New Hampshire campus. This was a two-month conference to which the leading lights of the newly emerging field were invited. The goals were both ambitious and optimistic; the conference proposal declared that "an attempt will be made to find how to make machines use language, form abstractions and concepts, solve kinds of problems now reserved for humans, and improve themselves" and promised that the organizers' believed a "significant advance can be made in one or more of these problems if a carefully selected group of scientists work on it together for a summer."[3] Attendees included Marvin Minsky, who along with McCarthy became one of the world's most celebrated AI researchers and founded the Computer Science and Artificial Intelligence Lab at MIT, and Claude Shannon, a legendary electrical engineer who formulated the principles of information theory that underlie electronic communication and make the internet possible.

The brightest mind, however, was notably absent from the Dartmouth conference. Alan Turing had committed suicide two years earlier. Prosecuted for a same-sex relationship under the "indecency" laws then in force in Britain, Turing was given a choice between imprisonment or chemical castration through the forced introduction of estrogen. Depressed after selecting the second option, he took his own life in 1954. For the emergent fields of computer science and artificial intelligence, the loss would be incalculable. Turing was only forty-one when he died. In a more just world, he would have almost certainly lived to see the advent of the personal computer and quite possibly the rise of the internet, and perhaps even many of the innovations that followed. No one can say what contributions Turing might have

made over those decades, or how much further along the field
of artificial intelligence might now be, but the intellectual loss to
the field, and to all of humanity, is likely staggering.

The field of artificial intelligence made rapid progress in the
years following the Dartmouth conference. Computers were becoming more capable, important breakthroughs were made and
algorithms were developed that could solve an increasing range
of problems. Artificial intelligence as a field of study was introduced at universities across the United States, and a number of
AI research labs were established.

One of the most important enablers of this progress was
massive investment from the U.S. government, especially the
Pentagon. Much of this was funneled through the Advanced
Research Projects Agency, or ARPA. One especially important
center of APRA-funded research was the Stanford Research Institute, which later separated from Stanford University to become
SRI International. SRI's Artificial Intelligence Center, established
in 1966, did groundbreaking work in areas like language translation and speech recognition. The lab also created the first truly
autonomous robot, a machine capable of turning AI-powered
reasoning into physical interaction with the environment. Nearly
half a century after its founding, SRI's Artificial Intelligence Center would spin off a startup company with a new personal assistant called Siri that would be acquired by Apple in 2010.

Progress, however, soon led to overexuberance, outsized
promises and unrealistic expectations. In 1970, *LIFE* published
an article about the robot developed at SRI, calling it the world's
first "electronic person." Marvin Minsky, who was then a star
AI researcher at MIT, declared to the article's author, Brad
Darrach, with "quiet certitude":

> In from three to eight years we will have a machine with
> the general intelligence of an average human being. I mean

a machine that will be able to read Shakespeare, grease a car, play office politics, tell a joke, have a fight. At that point the machine will begin to educate itself with fantastic speed. In a few months it will be at genius level and a few months after that its powers will be incalculable.[4]

Darrach checked this statement with other AI researchers and was told that perhaps Minsky's three-to-eight-year time frame was a bit optimistic. It might take fifteen years, they said, but "all agreed that there would be such a machine and that it could precipitate the Third Industrial Revolution, wipe out war and poverty, and roll up centuries of growth in science, education and the arts."[5]

As it became evident that such predictions were wildly off the mark, and that building AI systems capable of performing even far less ambitious tasks was proving to be much more difficult than expected, enthusiasm began to drain away from the field. By 1974, disillusionment among investors, especially the government agencies that played an outsized funding role, cast a pall over the field—and over the career prospects of many AI researchers. Throughout its history, the field of artificial intelligence has suffered from a kind of collective bipolar disorder, with periods of high exuberance and rapid progress punctuated by sometimes decades-long stretches of disillusionment and low investment that have come to be called "AI winters."

The field's periodic plunge into AI winters likely resulted in part from a lack of appreciation of just how difficult the problems that AI aims to solve really are. Another critical factor, however, was a simple failure to recognize how genuinely slow computers were prior to the 1990s. It would take decades of progress under the relentless regime of Moore's Law to deliver hardware that would begin to put the dreams of the 1956 Dartmouth conference participants within reach.

The arrival of that faster computer hardware led to some dramatic advances in the late 1990s. In May 1997, IBM's Deep Blue computer narrowly defeated the world chess champion, Garry Kasparov, in a six-game match. Though this was generally heralded as a triumph for artificial intelligence, in fact it was primarily a feat accomplished by leveraging the power of brute computation. The specialized algorithms that ran on Deep Blue's refrigerator-sized, custom-designed hardware were able to look far ahead, rapidly sifting through a multitude of potential moves in a way that would have been impossible for even the most capable human mind.

IBM triumphed again in 2011 with the advent of Watson, a machine that easily defeated the world's top contestants on the TV game show *Jeopardy!* In many ways this was a far more impressive feat because it required an understanding of natural language that included even the ability to navigate jokes and puns. Unlike Deep Blue, Watson was a system that could go beyond the confines of a board game with rigidly defined rules and deal with a seemingly limitless body of information. Watson won at *Jeopardy!* by simultaneously deploying a swarm of smart algorithms that coursed through reams of data, often drawn from Wikipedia articles, to determine the correct responses as it played the game.

Watson heralded a new age and portended machines that would finally begin to parse language and truly engage with humans, but 2011 would also mark the beginning of a dramatic shift in the underlying technology of artificial intelligence. Watson relied on machine learning algorithms that used statistical techniques to make sense of information, but over the next few years, another kind of machine learning—based directly on the perceptron conceived by Frank Rosenblatt more than half a century earlier—would once again come to the forefront and then rapidly rise to dominate the field of artificial intelligence.

CONNECTIONIST VS. SYMBOLIC AI
AND THE RISE OF DEEP LEARNING

Even as the general field of artificial intelligence traced its boom-and-bust path over the decades, the research focus swung between two general philosophies that emphasized contrasting approaches to building more intelligent machines. One school of thought grew out of Rosenblatt's work on neural networks in the 1950s. Its adherents believed that an intelligent system should be modeled on the fundamental architecture of the brain and that it should utilize deeply connected components based loosely on biological neurons. This approach, which came to be called "connectionism," emphasized learning as the central capability of intelligence and argued that if a machine could be made to efficiently learn from data, then the other capabilities exhibited by the human brain might eventually emerge as well. There was, after all, strong evidence for the effectiveness of this model: the human brain itself was known to be composed entirely of an incomprehensibly complex system of interconnected biological neurons.

The competing camp consisted of researchers who embraced a "symbolic" approach that emphasized the application of logic and reasoning. For the symbolists, learning was not so important. Instead, the key to intelligence was the ability to leverage knowledge through reasoning, decision-making and action. Rather than designing algorithms that could learn by themselves, symbolists manually encoded information directly into the systems they built, a practice that gave rise to a field known as "knowledge engineering."

Symbolic AI was the engine that powered nearly all early practical applications of artificial intelligence. Knowledge engineers, working with doctors, for example, were able to build systems that attempted to diagnose illnesses using algorithms that employed decision trees. Such medical expert systems produced mixed results and often proved to be inflexible and unreliable. However, in many other applications, such as the autopilot systems used on jet aircraft, the techniques that were developed through research into expert systems have gradually become routine components of software design and are no longer labeled as "artificial intelligence."

Connectionism had its origins in research that aimed to understand the function of the human brain. In the 1940s, Warren McCulloch and Walter Pitts introduced the idea of an artificial neural network as a kind of computational approximation for the way the biological neurons in the brain operated.[6] Frank Rosenblatt, who was trained as a psychologist and lectured in the psychology department at Cornell, later incorporated these ideas into his perceptron.

The perceptron was capable of rudimentary pattern recognition tasks like recognizing printed characters via a camera that was attached to the device. The inventor and author Ray Kurzweil, who is now an engineering director at Google, met Rosenblatt in his lab at Cornell in 1962. Kurzweil told me that he brought samples to try out on the perceptron and that the machine worked perfectly as long as the characters were printed clearly in the proper font. Rosenblatt told the young Kurzweil, who was then about to matriculate at MIT, that he was confident that much better results could be obtained if perceptrons were cascaded into multiple levels, with the output of one level feeding into the inputs of the next.[7] Rosenblatt, however, died in a boating accident in 1971, having never built a multilayer implementation.

The initial enthusiasm for artificial neural networks began to drain away by the late 1960s. One of the most important drivers of this decline was the publication of the 1969 book *Perceptrons*, co-authored by Marvin Minsky. While Minsky was extraordinarily confident about the prospects for artificial intelligence as a whole, he was ironically very pessimistic about the specific approach that would one day lead to unprecedented progress. In the book, Minsky and co-author Seymour Papert presented formal mathematical proofs that highlighted the limitations of neural networks and suggested that the technology would prove unable to solve sophisticated practical problems.[8]

As computer scientists and graduate students began to shy away from working on neural networks, the symbolic AI approach—now often referred to as "classical AI"—became dominant. Neural networks would see brief resurgences in the 1980s and again in the 1990s, but for decades the symbolic school would reign, even as enthusiasm for the field of artificial intelligence as a whole cycled between extremes. For the connectionists, AI winters unfolded with frightening severity and duration, often persisting even while the practitioners of symbolic AI enjoyed a balmy spring.

Throughout the 1970s and early 1980s, the freeze was especially hard. Yann LeCun, who is now considered to be one of the primary architects of deep learning, told me that during this period, research into neural networks was "worse than marginalized" and that "you couldn't publish a paper that even mentioned the phrase 'neural networks' because it would immediately be rejected."[9] Still, a small number of researchers maintained faith in the connectionist vision. Many of these individuals had backgrounds not in computer science but in psychology or human cognition, and their interest was driven by a desire to create a mathematical model for the function of the brain. In the early 1980s, David Rumelhart, a psychology professor at the University of California, San Diego, conceived the technique known as "backpropagation," which is still the primary learning algorithm used in multilayered neural networks today. Rumelhart, along with Ronald Williams, a computer scientist at Northeastern University, and Geoffrey Hinton, then at Carnegie Mellon, described how the algorithm could be used in what is now considered to be one of the most important scientific papers in artificial intelligence, published in the journal *Nature* in 1986.[10] Backpropagation represented the fundamental conceptual breakthrough that would someday lead deep learning to dominate the field of AI, but it would be decades before computers would become fast enough to truly leverage the approach. Geoffrey Hinton, who had been a young postdoctoral researcher working with Rumelhart at UC San Diego in 1981,[11] would go on to become perhaps the most prominent figure in the deep learning revolution.

By the end of the 1980s, practical applications for neural networks began to emerge. Yann LeCun, then a researcher at AT&T's Bell Labs, used the backpropagation algorithm in a new architecture called a "convolutional neural network." In convolutional networks, the artificial neurons are connected in a way that is inspired by the visual cortex in the brains of mammals, and these networks were designed to be especially effective at image recognition. LeCun's system could recognize handwritten digits, and by the late 1990s convolutional neural networks were allowing ATM machines to understand the numbers written on bank checks.

The 2000s saw the rise of "big data." Organizations and governments were now collecting and attempting to analyze information at what would, just a short time ago, have been an unimaginable scale, and it had become evident that the total volume of data generated globally would continue to grow at an exponential pace. This data gusher would soon intersect with the latest machine learning algorithms to enable a revolution in artificial intelligence.

One of the most consequential new data troves resulted from the efforts of a young computer science professor at Princeton

University. Fei-Fei Li, whose work was focused on computer vision, realized that teaching machines to make visual sense of the real world would require a comprehensive teaching resource with properly labeled examples showing many variations of people, animals, buildings, vehicles, objects—and just about anything else one might encounter. Over a two-and-a-half-year period, she set out to give titles to more than three million images across over 5,000 categories. This work had to be done manually; only a human being could make the proper association between a photograph and a descriptive label. Because the cost of hiring even undergraduates to take on such a massive task would have been prohibitive, Li's team turned to Amazon's Mechanical Turk, a newly developed platform that crowdsourced information-oriented tasks to remote workers, often in low-wage countries.[12]

Li's project, known as ImageNet, was published in 2009 and soon became an indispensable resource for research into machine vision. Beginning in 2010, Li organized an annual competition in which teams from universities and corporate research labs turned their algorithms loose to try to label images drawn from the massive dataset. The ImageNet Large Scale Visual Recognition Competition held two years later, in September 2012, arguably represents the inflection point for the technology of deep learning.[13] Geoff Hinton, along with Ilya Sutskever and Alex Krizhevsky from his research lab at the University of Toronto, entered a many-layered convolutional neural network that dramatically outpaced competing algorithms and offered unambiguous evidence that deep neural networks had finally evolved into a genuinely practical technology. The triumph of Hinton's team resonated widely within the AI research community and put a spotlight on the productive coupling of massive datasets with powerful neural algorithms—a symbiosis that would soon produce advances that just a few years ago had been anchored firmly in the realm of science fiction.

The story I've sketched out here represents roughly what you might call the "standard history" of deep learning. In this telling, the 2018 Turing Award recipients Geoff Hinton, Yann LeCun and Yoshua Bengio, a professor at the University of Montreal, loom especially large—so much so that they are often referred to as the "godfathers of deep learning." (They are also sometimes called "the godfathers of AI," a vivid demonstration of the extent to which deep learning has come to completely dominate the field, pushing aside the earlier focus on symbolic approaches.) There are other versions of this history, however. As with most scientific fields, the competition for recognition is keen—and has perhaps been driven to extremes by the growing sense that progress in AI has crossed thresholds that will inevitably lead to genuinely historic transformations of both society and the economy.

The most vocal proponent of an alternate history is Jürgen Schmidhuber, who co-directs the Dalle Molle Institute for Artificial Intelligence Research in Lugano, Switzerland. In the 1990s, Schmidhuber and his students developed a special type of neural network that implemented "long short-term memory," or LSTM. With LSTM, networks are able to "remember" data from the past and incorporate it into the current analysis. This has proven to be of critical importance in areas like speech recognition and language translation, where the context created by words that came previously has a huge impact on accuracy. Companies like Google, Amazon and Facebook all rely heavily on LSTM, and Schmidhuber feels that it is the work of his team, rather than that of the more celebrated North American researchers, that underlies much of AI's recent progress.

In an email written to me shortly after the publication of my book *Architects of Intelligence*—in which I included a short summary of deep learning's standard history—Schmidhuber told me that "many of the things you wrote are quite misleading and, as a consequence, quite frustrating!"[14] Deep learning, according to Schmidhuber, has its roots not in the United States or Canada but rather in Europe. The first learning algorithms for multilayered neural networks, he says, were described by the Ukrainian researcher Alexey Grigorevich Ivakhnenko in 1965, while the backpropagation algorithm was published in 1970—a decade and a half before the famous paper by Rumelhart—by Finnish student Seppo Linnainmaa. Schmidhuber is clearly frustrated over the lack of recognition given to his own research, and is known for abrasively interrupting presentations at AI conferences and leveling accusations of a "conspiracy" to rewrite deep learning's history, especially on the part of Hinton, LeCun and Bengio.[15] For their part, these better-known researchers push back aggressively. LeCun told a *New York Times* reporter that "Jürgen is manically obsessed with recognition and keeps claiming credit he doesn't deserve."[16]

Though disagreements about the true origins of deep learning are likely to persist, there is no doubt that in the wake of the 2012 ImageNet competition, the technique rapidly took the field of artificial intelligence—and most of the technology industry's largest companies—by storm. American tech behemoths like Google, Amazon, Facebook and Apple, as well as the Chinese companies Baidu, Tencent and Alibaba, immediately recognized the disruptive potential of deep neural networks and began to build research teams and incorporate the technology into their products and operations. Google hired Geoff Hinton, Yann LeCun became the director of Facebook's new AI research lab, and the entire industry began waging a full-on talent war that pushed salaries and stock options for even newly minted graduate students with expertise in deep learning into the stratosphere. In 2017, CEO Sundar Pichai declared that Google was now an "AI-first company" and said that artificial intelligence would be one of the most important dimensions along which the company would compete with the other tech giants.[17] At Google and Facebook, the technology was deemed so important that deep learning researchers were assigned offices in direct proximity to the CEO,[18] and by the end of the decade, neural networks had so completely dominated the field that the media would often treat the terms "deep learning" and "artificial intelligence" as synonymous.

CHAPTER 5

DEEP LEARNING AND THE FUTURE OF ARTIFICIAL INTELLIGENCE

THE EMBRACE OF DEEP LEARNING BY THE WORLD'S LARGEST technology companies, together with the arrival of ever more compelling consumer and business applications that leverage the power of neural networks, leaves little doubt that the technology is here to stay. However, there is a growing sense that the rate of progress is not sustainable, and that future advances will require significant new innovations. As we'll see, one of the most important questions going forward will be whether or not the AI pendulum will once again swing back toward approaches that emphasize symbolic AI and, if so, how those ideas can be successfully integrated with neural networks. Before delving into the future of artificial intelligence, let's briefly look in more detail at just how a deep learning system actually works and how these networks are trained to perform useful tasks.

HOW DEEP NEURAL NETWORKS WORK

The media often describes deep learning systems as "brain-like," and this can easily lead to misconceptions about just how closely the neural networks used in artificial intelligence are modeled on their biological counterparts. The human brain is arguably the most complex system in the known universe, with roughly one hundred billion neurons and hundreds of trillions of interconnections. But this staggering level of complexity does not arise simply from connectivity at a massive scale; rather, it extends to the operation of the neurons themselves and the way they transmit signals and adapt to new information over time.

A biological neuron consists of three main parts: the cell body, where the nucleus resides; numerous filaments known as "dendrites" that carry incoming electrical signals; and a single, much longer and finer filament called an "axon," along which the neuron transmits an outgoing signal to other neurons. Both the dendrites and the axon typically branch extensively, with the dendrites sometimes receiving electrical stimulation from tens of thousands of other neurons. When the collective signals arriving through the dendrites excite the neuron, it in turn delivers an outgoing electrical charge, known as an action potential, through the axon. However, the brain's connections are not electrically hardwired. Instead, the axon from one neuron transmits a chemical signal to a dendrite of another across a junction known as a "synapse." These electrochemical functions are critical to the brain's operation and to its ability to learn and adapt, but are, in many cases, not well understood. The chemical dopamine, for example, which is associated with pleasure or reward, is a neurotransmitter that operates within the synaptic gap.

An artificial neural network brushes aside nearly all these details and attempts to create a rough mathematical sketch of the way neurons operate and connect. If the brain is the Mona Lisa, then the structures used in deep learning systems might, at best, be something akin to Lucy from *Peanuts*. The basic plan for artificial neurons was conceived in the 1940s, and in the decades since, work on these systems has largely been decoupled from brain science; the algorithms that power deep learning systems have been developed independently, often through experimentation and without any specific attempt to simulate what might actually be occurring in the human brain.

To visualize an artificial neuron, imagine a container with three or more incoming pipes, each of which delivers a stream of water. These pipes correspond roughly to dendrites in the biological neuron. There is also an axon pipe to carry an outgoing stream of water. If the level of water delivered by the incoming pipes reaches a certain threshold, the neuron "fires" then sends an outgoing stream of water through its axon pipe.

The key feature that turns this simple contraption into a useful computational device is a valve that is fitted to each of the incoming pipes, so that the flow of water through the pipe can be controlled. By adjusting these valves, it's possible to directly regulate the influence that other connected neurons have on this particular neuron. The process of training a neural network to perform some useful task is essentially a matter of adjusting these valves—called "weights"—until the network can correctly identify patterns.

In a deep neural network, a software simulation of artificial neurons that work more or less like these containers would be arranged in a series of layers, so that the output from one layer of neurons is connected to the inputs of neurons in the next layer. The connections between neurons in adjacent layers are often simply set randomly; alternatively, in a specific neural architecture, such as a convolutional network designed to recognize images, the neurons may be connected according to a more deliberate plan. Sophisticated neural networks can contain more than a hundred layers and millions of individual artificial neurons.

Once such a network has been configured, it can then be trained to perform specific tasks, such as image recognition or language translation. For example, in order to train a neural network to recognize handwritten digits, the individual pixels from a photograph of a written numeral would become the inputs for the first layer of neurons. The answer, or in other words the number corresponding to the written digit, would arrive at the outputs from the last layer of artificial neurons. Training the network to generate the correct answer is a process of inputting training examples and then adjusting all the weights in the network so that it gradually converges on the right answer. Once the weights have been optimized in this way, the network can then be deployed on new examples that are not included in the training set.

Tuning the weights so the network eventually succeeds in converging on the right answer nearly every time is where the famous backpropagation algorithm comes in. A complex deep learning system might have a billion or more connections between neurons, each of which has a weight that needs to be optimized. Backpropagation essentially allows all the weights in the network to be adjusted collectively, rather than one at a time, delivering a massive boost to computational efficiency.[1] During the training process, the output from the network is compared to the correct answer, and information that allows each weight to be adjusted accordingly propagates back through the layers of neurons. Without backpropagation, the deep learning revolution would not have been possible.

While all this sketches out the basic mechanics of configuring and training a neural network so that it will produce useful results, it still leaves unanswered the fundamental question: what exactly is actually happening within one of these systems as it churns through data and delivers answers with an accuracy that is often superhuman?

The short explanation is that, within the neural network, a representation of knowledge is being created, and the level of abstraction for this knowledge increases in subsequent layers of the network. This is easiest to understand for a network configured to recognize visual images. The network's comprehension of the image begins at the level of pixels. In subsequent neural layers, visual features such as edges, curves and textures are perceived. Deeper within the system, still more complex representations emerge. Eventually, the system's understanding is so definitive that it captures the full essence of the image in a way that allows the network to identify it—even when confronted with a huge number of alternatives.

A more complete answer to the question, however, would admit we don't really know exactly what's happening—or at least we can't readily describe it. No programmer sets out to define the various levels of abstraction or the way in which knowledge is represented within the network. All of this emerges organically, and the representation is distributed across the millions of interconnected artificial neurons firing throughout the system. We know that the network in some sense comprehends the image, and yet it is very difficult, or even impossible, to accurately describe just what is coalescing within its neurons—and this becomes ever more the case as we progress deeper into the layers of the network, or if we examine systems that operate on types of data that are not so easily visualized. This relative opacity—the concern that deep neural networks in effect are "black boxes"— is one of the most important concerns that we will return to in Chapter 8.

The overwhelming majority of deep learning systems are trained to do useful tasks by presenting the network with a massive dataset that has been carefully labeled or categorized. For example, a deep neural network might be trained to correctly identify animals in photographs by being provided thousands, or perhaps even millions, of images, each correctly labeled with the name of the animal depicted. This training regimen, known as "supervised learning," can take many hours even when very high performance hardware is used.

Supervised learning is the training method used in perhaps ninety-five percent of practical machine learning applications. The technique powers AI radiology systems (trained with a huge number of medical images labeled either "Cancer" or "No Cancer"), language translation (trained with millions of documents pre-translated into different languages) and a nearly limitless number of other applications that essentially involve comparing and classifying different forms of information. Supervised learning typically requires vast amounts of labeled data, but the results can be very impressive—routinely resulting in systems with a superhuman ability to recognize patterns. Five years after the 2012 ImageNet competition that marked the onset of the deep learning explosion, the image recognition algorithms had become so proficient that the annual competition was reoriented toward a new challenge involving the recognition of real-world three-dimensional objects.[2]

In cases where labeling all this data requires the kind of interpretation that only a human can provide, as in attaching descriptive annotations to photographs, the process is expensive and cumbersome. One common solution is to copy the approach employed by Fei-Fei Li for the ImageNet dataset and turn to crowdsourcing. Platforms like Mechanical Turk make it possible to pay a distributed team of people pennies to do this work. The opportunity to streamline this process has given rise to a number of startup companies that are specifically focused on finding efficient ways to annotate data in preparation for supervised learning. The critical importance of accurately labeling massive datasets, especially for applications that involve understanding visual information, is especially well demonstrated by the meteoric ascent of Scale AI, which was founded by nineteen-year-old MIT dropout Alexandr Wang in 2016. Scale AI contracts with over 30,000 crowdsourced workers who label data for clients including Uber, Lyft, Airbnb and Alphabet's self-driving car division, Waymo. The company has received more than $100 million in venture capital and now ranks as a Silicon Valley "unicorn"—a startup valued in excess of $1 billion.[3]

In many other cases, however, nearly incomprehensible quantities of beautifully labeled data are generated seemingly automatically—and for the companies that possess it, virtually free of charge. The massive torrent of data generated by platforms like Facebook, Google or Twitter is valuable in large measure because it is carefully annotated by the people using the platforms. Every time you "like" or "retweet" a post, every time you view or scroll down a webpage, every video you watch (and the amount of time you spend watching it) and every time you undertake a myriad of other online actions, you in effect attach a label to some particular item of data. You—along with the millions of other people using one of the major platforms—are essentially stepping into the shoes of all those crowdsourced workers deployed by companies like Scale AI. It is, of course, not coincidental that most important AI research initiatives tend to be associated with large internet companies. The synergy between artificial intelligence and ownership of huge troves of data is often remarked on, but a critical factor underlying this symbiosis is the possession of a massive machine for annotating all that data at little or no cost, so it can then become fodder for a supervised learning regimen deployed on a powerful neural network.

While supervised learning dominates, another important technique—"reinforcement learning"—is used in certain applications. Reinforcement learning builds competence through repeated practice or trial and error. When an algorithm ultimately succeeds at a specified objective, it receives a digital reward. This is essentially the way a dog is trained. The animal's behavior may be random at first, but when it manages to sit in response to the proper command, it gets a treat. Repeat this process enough times and the dog will learn to reliably sit.

The leader in reinforcement learning is the London-based company DeepMind, which is now owned by Google's parent, Alphabet. DeepMind has made massive investments in research based on the technique, merging it with powerful convolutional neural networks to develop what the company calls "deep reinforcement learning." DeepMind began working on applying reinforcement learning to build AI systems that could play video games shortly after its founding in 2010. In January 2013, the company announced that it had created a system called DQN that was capable of playing classic Atari games, including *Space Invaders*, *Pong* and *Breakout*. DeepMind's system was able to teach itself to play the games by using only raw pixels and the game score as the learning inputs. After honing its technique on many thousands of simulated games, DQN turned in the best scores ever achieved by a computer for six of the games, and was able to defeat the best human players in three.[4] By 2015, the system had conquered forty-nine Atari games, and DeepMind declared that it had developed the first AI system that bridged "the divide between high-dimensional sensory inputs and actions" and that the DQN was "capable of learning to excel at a diverse array of challenging tasks."[5] These achievements caught the attention of Silicon Valley's titans, including especially Google founder Larry Page, and in 2014 Google brushed aside a competing offer from Facebook and acquired DeepMind for $400 million.

Deep reinforcement learning's most notable achievement came in March 2016, when AlphaGo, a system developed by DeepMind to play the ancient game of Go, defeated Lee Sedol, then one of the world's best players, in a five-game match in Seoul, South Korea. Proficiency at Go is held in extremely high regard in Asia, where the game has been played for millennia. Confucius's writings reference the game, and its origins may stretch back nearly to the dawn of Chinese civilization. According to one theory, Go was invented during the reign of the emperor Yao, sometime prior to 2000 BC.[6] The ability to play Go, along with expertise in calligraphy, painting and playing a stringed musical instrument, was viewed as one of the four primary arts that marked ancient Chinese scholarship.

Unlike chess, the game of Go is so complex as to be immune to the onslaught of brute force algorithms. During the course of the game, the board, which consists of a nineteen-by-nineteen grid, is largely filled with black and white game pieces called "stones." As DeepMind CEO Demis Hassabis often likes to point out when he discusses AlphaGo's accomplishment, the number of possible arrangements of the stones on the board exceeds the estimated number of atoms in the known universe. Over the thousands of years that the game has been played, it is extraordinarily—indeed vanishingly—unlikely that any two games have unfolded in identical fashion. In other words, any attempt to look ahead and account for the entire range of possible future moves, as might be done for a game with more confining rules, is computationally out of reach—even for the most powerful hardware.

Aside from this vast level of complexity, it seems evident that playing Go draws heavily on what might be called human intuition. The best players are often at a loss when they are asked to explain exactly why they chose a particular strategy. Instead, they might describe a "feeling" that led them to place a stone at a particular point on the board. This is precisely the type of undertaking that seems as though it ought to be beyond the capability of a computer—a job that we can rightfully expect to be safe from the threat of automation, at least for the foreseeable future. Nonetheless, the game of Go fell to the machines at least a decade before most computer scientists believed such a feat would be possible.

The DeepMind team began by using a supervised learning technique to train AlphaGo's neural networks on thirty million moves extracted from detailed records of games played by the best human players. It then turned to reinforcement learning, essentially turning the system loose to play against itself. Over the course of thousands of simulated practice games, and under the relentless pressure of a reward-based drive to improve, AlphaGo's deep neural networks gradually progressed toward superhuman proficiency.[7] The triumph of AlphaGo over Lee Sedol in 2016, and then over the world's top-ranked player, Ke Jie, a year later, once again sent shock waves through the AI research community. The achievements may also have led to what venture capitalist and author Kai-Fu Lee has called a "Sputnik moment" in China—in the wake of which the government quickly moved to position the country to become a leader in artificial intelligence.[8]

While supervised learning depends on massive quantities of labeled data, reinforcement learning requires a huge number of practice runs, the majority of which end in spectacular failure. Reinforcement learning is especially well suited to games—in which an algorithm can rapidly churn though more matches than a human being could play in a lifetime. The approach can also be applied to real-world activities that can be simulated at high speed. The most important practical application of reinforcement learning is currently in training self-driving cars. Before the autonomous driving systems used by Waymo or Tesla ever see a real car or a real road, they are trained at high speed on powerful computers, through which the simulated cars gradually learn after suffering catastrophic crashes thousands of times. Once the algorithms are trained and the crashes are in the past, the software can then be transferred to real-world cars. Though this approach is generally effective, it goes without saying that no license-seeking sixteen year old needs to crash a thousand times before figuring out how to drive. This stark contrast between how learning works in machines and how it operates with vastly less data in the human brain highlights both the limitations of current AI systems and the enormous potential for the technology to improve going forward.

WARNING SIGNS

The 2010s were arguably the most exciting and consequential decade in the history of artificial intelligence. Though there have certainly been conceptual improvements in the algorithms used in AI, the primary driver of all this progress has simply been deploying more expansive deep neural networks on ever faster computer hardware where they can hoover up greater and greater quantities of training data. This "scaling" strategy has been explicit since the 2012 ImageNet competition that set off the deep learning revolution. In November of that year, a front-page *New York Times* article was instrumental in bringing awareness of deep learning technology to the broader public sphere. The article, written by reporter John Markoff, ends with a quote from Geoff Hinton: "The point about this approach is that it scales beautifully. Basically you just need to keep making it bigger and faster, and it will get better. There's no looking back now."[9]

There is increasing evidence, however, that this primary engine of progress is beginning to sputter out. According to one analysis by the research organization OpenAI, the computational resources required for cutting-edge AI projects is "increasing exponentially" and doubling about every 3.4 months.[10]

In a December 2019 *Wired* magazine interview, Jerome Pesenti, Facebook's Vice President of AI, suggested that even for a company with pockets as deep as Facebook's, this would be financially unsustainable:

> When you scale deep learning, it tends to behave better and to be able to solve a broader task in a better way. So, there's an advantage to scaling. But clearly the rate of progress is not sustainable. If you look at top experiments, each year the cost [is] going up 10-fold. Right now, an experiment might be in seven figures, but it's not going to go to nine or ten figures, it's not possible, nobody can afford that.[11]

Pesenti goes on to offer a stark warning about the potential for scaling to continue to be the primary driver of progress: "At some point we're going to hit the wall. In many ways we already have." Beyond the financial limits of scaling to ever larger neural networks, there are also important environmental considerations. A 2019 analysis by researchers at the University of Massachusetts, Amherst, found that training a very large deep learning system could potentially emit as much carbon dioxide as five cars over their full operational lifetimes.[12]

Even if the financial and environmental impact challenges can be overcome—perhaps through the development of vastly more efficient hardware or software—scaling as a strategy simply may not be sufficient to produce sustained progress. Ever-increasing investments in computation have produced systems with extraordinary proficiency in narrow domains, but it is becoming increasingly clear that deep neural networks are subject to reliability limitations that may make the technology unsuitable for many mission critical applications unless important conceptual breakthroughs are made. One of the most notable demonstrations of the technology's weaknesses came when a group of researchers at Vicarious—the small company focused on building dexterous robots that we met in Chapter 3—performed an analysis of the neural network used in Deep-Mind's DQN, the system that had learned to dominate Atari video games.[13] One test was performed on *Breakout*, a game in which the player has to manipulate a paddle to intercept a fast-moving ball. When the paddle was shifted just a few pixels higher on the screen—a change that might not even be noticed by a human player—the system's previously superhuman performance immediately took a nose dive. DeepMind's software had no ability to adapt to even this small alteration. The only way to get back to top-level performance would have been to start from scratch and completely retrain the system with data based on the new screen configuration.

What this tells us is that while DeepMind's powerful neural networks do instantiate a representation of the *Breakout* screen, this representation remains firmly anchored to raw pixels even at the higher levels of abstraction deep in the network. There is clearly no emergent understanding of the paddle as an actual object that can be moved. In other words, there is nothing close to a human-like comprehension of the material objects that the pixels on the screen represent or the physics that govern their movement. It's just pixels all the way down. While some AI researchers may continue to believe that a more comprehensive understanding might eventually emerge if only there were more layers of artificial neurons, running on faster hardware and consuming still more data, I think this is very unlikely. More fundamental innovations will be required before we begin to see machines with a more human-like conception of the world.

This general type of problem, in which an AI system is inflexible and unable to adapt to even small unexpected changes in its input data, is referred to, among researchers, as "brittleness." A brittle AI application may not be a huge problem if it results in a warehouse robot occasionally packing the wrong item into a box. In other applications, however, the same technical shortfall can be catastrophic. This explains, for example, why progress toward fully autonomous self-driving cars has not lived up to some of the more exuberant early predictions.

As these limitations came into focus toward the end of the decade, there was a gnawing fear that the field had once again gotten over its skis and that the hype cycle had driven expectations to unrealistic levels. In the tech media and on social media, one of the most terrifying phrases in the field of artificial intelligence—"AI winter"—was making a reappearance. In a January 2020 interview with the BBC, Yoshua Bengio said that "AI's abilities were somewhat overhyped . . . by certain companies with an interest in doing so."[14]

A large share of this concern came to bear on the industry that was, as we saw in Chapter 3, at the absolute summit of all the accumulated hype: self-driving cars. It was becoming clear that, despite optimistic predictions early in the decade, truly driverless vehicles, capable of navigating in a wide range of conditions, were still not close to reality. Companies like Waymo, Uber and Tesla had put autonomous vehicles on public roads, but outside a few very constrained experiments, there was always a human driver—who, it turned out, had to take control of the car all too often. Even with a driver in place to oversee the car's operation, a number of fatal accidents had tarnished the industry's reputation. In a widely shared 2018 blog post entitled "AI Winter Is Well on Its Way," machine learning researcher Filip Piekniewski pointed out that records required by the State of California showed that one car being tested "literally could not drive ten miles" without a system disengagement that required the human driver to take control.[15]

My own view is that if another AI winter indeed looms, it's likely to be a mild one. Though the concerns about slowing progress are well founded, it remains true that over the past few years AI has been deeply integrated into the infrastructure and business models of the largest technology companies. These companies have seen significant returns on their massive investments in computing resources and AI talent, and they now view artificial intelligence as absolutely critical to their ability to compete in the marketplace. Likewise, nearly every technology startup is now, to some degree, investing in AI, and companies large and small in other industries are beginning to deploy the technology. This successful integration into the commercial sphere is vastly more significant than anything that existed in prior AI winters, and as a result the field benefits from an army of advocates throughout the corporate world and has a general momentum that will act to moderate any downturn.

There's also a sense in which the fall of scalability as the primary driver of progress may have a bright side. When there is a widespread belief that simply throwing more computing resources at a problem will produce important advances, there is significantly less incentive to invest in the much more difficult work of true innovation. This was arguably the case, for example, with Moore's Law. When there was near absolute confidence that computer speeds would double roughly every two years, the semiconductor industry tended to focus on cranking out ever faster versions of the same microprocessor designs from companies like Intel and Motorola. In recent years, the acceleration in raw computer speeds has become less reliable, and our traditional definition of Moore's Law is approaching its end game as the dimensions of the circuits imprinted on chips shrink to nearly atomic size. This has forced engineers to engage in more "out of the box" thinking, resulting in innovations such as software designed for massively parallel computing and entirely new chip architectures—many of which are optimized for the complex calculations required by deep neural networks. I think we can expect the same sort of idea explosion to happen in deep learning, and artificial intelligence more broadly, as the crutch of simply scaling to larger neural networks becomes a less viable path to progress.

THE QUEST FOR MORE
GENERAL MACHINE INTELLIGENCE

Overcoming the current limitations of deep learning systems will require innovations that bring machine intelligence inexorably closer to the capabilities of the human brain. There are many significant obstacles along this path, but it culminates in what has always been the Holy Grail of artificial intelligence: a machine that can communicate, reason and conceive new ideas at the level of a human being or beyond. Researchers often refer to this as "artificial general intelligence," or AGI. Nothing close to AGI currently exists in the real world, but there are many examples from science fiction—including HAL from *2001: A Space Odyssey*, the Enterprise main computer and Mr. Data from *Star Trek* and, of course, the truly dystopian technologies portrayed in *The Terminator* and *The Matrix*. One can make a strong argument that the development of general machine intelligence with superhuman capability would be the most consequential innovation in the history of humanity: such a technology would become the ultimate intellectual tool, dramatically accelerating the rate of progress in countless areas. Among AI experts, opinions vary widely as to just how long it might take to achieve AGI. A few researchers are extremely optimistic, suggesting this breakthrough could occur within five to ten years. Others are far more cautious and believe that it might take a hundred years or more.

For the foreseeable future, most research is focused not so much on the actual achievement of human-level AI, but rather on the journey toward it and the numerous important innovations that will be required to successfully navigate the obstacles along the way. The quest to build a true thinking machine is not just a speculative science project; rather, it represents a kind of road map toward building AI systems that overcome current limitations and exhibit new capabilities. Progress along that path is nearly certain to spawn a wealth of practical applications with enormous commercial and scientific value.

This coupling of practical near-term innovation with the far more aspirational quest for true human-level machine intelligence is demonstrated by the research philosophies of the various teams working on AI at Google. Jeff Dean, Google's overall director of artificial intelligence, told me that while DeepMind, the independent company Google acquired in 2014, is specifically oriented toward general machine intelligence with a "structured plan" to solve specific issues in the hope of eventually achieving AGI, other research groups at Google take a "more organic" approach, with a focus on doing things "that we know are important but that we can't do yet, and once we solve those, then we figure out what is the next set of problems that we want to solve." All the AI research groups at Google, he says, are "working together on trying to build really intelligent, flexible AI systems."[16] Only time will tell whether a top-down planned approach or a process of step-by-step exploration will be more successful, but both paths will in all likelihood generate important new ideas with immediate applications.

Progress along these paths is being led by teams with varying research philosophies and many different strategies for confronting the challenges that lie ahead. What they all have in common is that their ultimate objectives are modeled on capabilities that, at least so far, are exclusive to human cognition.

One important approach is to look directly to the inner workings of the human brain for inspiration. These researchers believe that artificial intelligence should be directly informed by neuroscience. The leader in this area is DeepMind. The company's founder and CEO, Demis Hassabis—unusually for an AI researcher—received his graduate training in neuroscience, rather than computing, and holds a PhD in the field from University College, London. Hassabis told me that the single largest research group at DeepMind consists of neuroscientists who are focused on finding ways to apply the latest insights from brain science to artificial intelligence.[17]

Their objective is not to replicate the way the brain works in any detailed sense, but rather to be inspired by the basic principles that underlie its operation. AI experts often explain this approach by using an analogy to the achievement of powered flight and the subsequent development of modern aircraft designs. While airplanes are clearly inspired by birds, they do not, of course, flap their wings or otherwise attempt to directly mimic avian flight. Rather, once engineers understood the science of aerodynamics, it became possible to build machines that operate according to the same fundamental principles that allow birds to fly, but which are in most ways far more capable than their biological counterparts. Hassabis and the team at DeepMind believe that there might be a kind of "aerodynamics of intelligence"—a foundational theory that underlies human and, potentially, machine intelligence.

DeepMind's cross-disciplinary team delivered some compelling evidence that such a general set of principles might indeed exist when the company published research in May 2018. Four years earlier, the Nobel Prize in Physiology or Medicine had been awarded to three neuroscientists, John O'Keefe, May-Britt Moser and Edvard Moser, for their discovery of a special type of neuron that enables spatial navigation in animals. These neurons, called grid cells, fire in a regular hexagonal pattern within the brain as the animal explores its environment. Grid cells are thought to make up a kind of "internal GPS," a neural representation of a mapping system that allows animals to remain oriented as they find their way through complex and unpredictable environments.

DeepMind conducted a computational experiment in which the company's researchers trained a powerful neural network on data that simulated the kind of movement-based information that an animal foraging in the dark might rely on. Remarkably, the researchers found that grid cell–like structures "spontaneously emerged within the network—providing a striking convergence with the neural activity patterns observed in foraging mammals."[18] In other words, the same fundamental navigation structure appears to arise naturally in two entirely different substrates, one biological and the other digital. Hassabis told me he considers this to be one of the company's most important breakthroughs and that the research may indicate that an internal system utilizing grid cells may simply be the most computationally efficient way to represent navigation information in any system, regardless of the details of its implementation.[19] DeepMind's scientific paper describing the research, published in the journal *Nature*,[20] resonated widely within the field of neuroscience, and insights like this suggest the company's interdisciplinary approach will likely turn out to be a two-way street, with AI research not only drawing upon lessons from the brain but also contributing to its understanding.

DeepMind once again made an important contribution to neuroscience when the company leveraged its expertise in reinforcement learning in early 2020 by exploring the operation of dopamine neurons in the brain.[21] Since the 1990s, neuroscientists have understood that these special neurons make a prediction about the likely reward that will result when an animal takes a specific action. If it turns out that the reward actually achieved is greater than expected, then relatively more dopamine is released. If the result underperforms, then less of this "feel good" chemical is generated. Computational reinforcement learning traditionally works in much the same way; the algorithm makes a prediction and then adjusts the reward based on the difference between the predicted and actual results.

Researchers at DeepMind were able to greatly improve a reinforcement learning algorithm by generating a distribution of predictions, rather than a single average prediction, and then adjusting the rewards accordingly. The company then teamed with a group of researchers at Harvard to see if the same kind of thing might be happening within the brain. They were able to show that mouse brains actually employ a similar distribution of predictions, with some dopamine neurons being relatively more pessimistic and others more optimistic about the potential reward. In other words, the company had once again demonstrated the same fundamental mechanism achieving parallel results in both a digital algorithm and the biological brain.

Research of this type reflects the confidence that Hassabis and his team have in reinforcement learning and their belief that it is a critical component of any attempt to progress toward more general artificial intelligence. In this, they are something of an outlier. Facebook's Yann LeCun, for example, has stated that he believes reinforcement learning plays a relatively minor role. In his presentations, he often says that if intelligence were a black forest cake, then reinforcement learning would amount to only the cherry on top.[22] The team at DeepMind believes it is far more central—and that it possibly provides a viable path to achieving AGI.

We generally describe reinforcement learning in terms of a reward-driven algorithm to optimize some external macro process—for example, learning to play the game of Go or figuring out how to drive a simulated car. However, Hassabis points out that reinforcement learning also plays a critical role internal to the brain, and that it may be essential to the emergence of intelligence. It's conceivable that reinforcement learning might be the primary mechanism that drives the brain toward curiosity, learning and reason. Imagine, for example, that the brain's inherent objective is simply to explore and then bring order to the torrent of raw data constantly bombarding an animal as it moves through its environment. Hassabis says that "we know seeing novel things releases dopamine in the brain" and if the brain is wired so that "finding information and structure [is] rewarding in itself, then that's a highly useful motivation."[23] In other words, the engine powering our continuous drive to understand the world around us could be a reinforcement learning algorithm linked to the production of dopamine.

An entirely different approach to building more general machine intelligence is being pursued by David Ferrucci, the CEO and founder of the AI startup company Elemental Cognition. Ferrucci is best known for leading the team that created IBM's Watson, the system that defeated Ken Jennings and other top *Jeopardy!* contestants in 2011. After Watson's triumph, Ferrucci left IBM and joined the Wall Street hedge fund Bridgewater and Associates, where he reportedly worked on using artificial intelligence to make sense of the macroeconomy and helped build Bridgewater founder Ray Dalio's management and investment philosophies into algorithms deployed throughout the firm.

Ferrucci now splits his time between his position as director of applied AI at Bridgewater and running Elemental Cognition, which received initial venture funding from the hedge fund.[24] Ferrucci told me that Elemental Cognition is focused on "real language understanding." The company is building algorithms that can autonomously read text and then engage in interactive dialog with humans in order to enhance the system's understanding of the material and also explain any conclusions. Ferrucci goes on to say:

> We want to look beyond the surface structure of language, beyond the patterns that appear in word frequencies, and get at the underlying meaning. From that, we want to be able to build the internal logical models that humans would create and use to reason and communicate. We want to ensure a system that produces a compatible intelligence. That compatible intelligence can autonomously learn and refine its understanding through human interaction, language, dialog, and other related experiences.[25]

This is an extraordinarily ambitious goal that, to me, sounds very close to human-level intelligence. Existing AI systems that process natural language suffer from a similar limitation to the one we saw with DeepMind's Atari-playing DQN when the game paddle was shifted a few pixels higher. Just as DQN has no understanding that the pixels on the screen represent a physical object that can be moved, current language systems have no real comprehension of what the words they process mean. This is the challenge that Elemental Cognition is taking on.

Ferrucci clearly believes that solving the language understanding problem represents the clearest path to more general intelligence. Rather than delving into the physiology of the brain in the way that DeepMind's team is attempting, Ferrucci argues that it is possible to directly engineer a system that can approach human level in its comprehension of language and its ability to employ logic and reason. He is unusual among AI researchers in that he feels the basic building blocks for general intelligence are already in place, or as he put it, "I don't think, as other people might, that we don't know how to do it and we're waiting for some enormous breakthrough. I don't think that's the case; I think we do know how to do it, we just need to prove that."[26]

He's also very optimistic about the prospects for achieving this goal in the relatively near future. In a 2018 documentary film, he said, "In three to five years, we'll have a computer system that can autonomously learn to understand and how to build understanding, not unlike the way a human mind works."[27] When I pressed him on this prediction, he backed off somewhat, acknowledging that that three to five years might indeed be optimistic. However, he said he would still "argue that it's something that we could see within the next decade or so. It's not going to be a 50- or a 100-year wait."[28]

To accomplish this goal, the team at Elemental Cognition is building a kind of hybrid system that includes deep neural networks as well as other machine learning approaches in combination with software modules built using traditional programming techniques to handle logic and reasoning. As we will see, the debate over the efficacy of such a hybrid approach as opposed to a strategy based entirely on neural networks is emerging as one of the most important questions confronting the field of AI.

Ray Kurzweil, now a director of engineering at Google, is likewise pursuing general intelligence along a path that is heavily oriented toward understanding language. Kurzweil is famous for his 2005 book *The Singularity Is Near*,[29] which established him as the most prominent evangelist for the idea of the "Singularity." Kurzweil and his many followers believe that the Singularity, likely brought on by the advent of superhuman machine intelligence, will someday mark an abrupt upward bend in humanity's historical arc—an inflection point when technological acceleration will become so extreme as to completely, and perhaps incomprehensibly, transform every aspect of human life and civilization.

In 2012, Kurzweil published another book entitled *How to Create a Mind*, in which he sketched out a conceptual model for human cognition.[30] According to Kurzweil, the brain is powered by around 300 million hierarchical modules, each of which "can recognize a sequential pattern and accept a certain amount of variability."[31] Kurzweil believes that this modular approach will ultimately result in a system that can learn from far less data than is the case with current deep learning systems that rely on supervised or reinforcement learning techniques. When Kurzweil approached Google's Larry Page to seek funding for a venture to put these ideas into practice, Page instead convinced him to come to Google and pursue his vision by taking advantage of the company's enormous computing resources.

Kurzweil has predicted for decades, and still believes, that AGI will be achieved sometime around the year 2029. Unlike many AI researchers, he continues to have faith in the Turing test as an effective measure of human-level intelligence. Conceived by Alan Turing in his 1950 paper, the test essentially amounts to a chat session in which a judge attempts to determine if the conversers are human or machine. If the judge, or perhaps a panel of judges, cannot distinguish the computer from a human, then the computer is said to pass the Turing test. Many experts are dismissive of the Turing test as an effective measure of human-level machine intelligence, in part because it has proven to be susceptible to gimmicks. In 2014, for example, in a contest held at the University of Reading in the United Kingdom, a chatbot that emulated a thirteen-year-old Ukrainian boy managed to fool the judges into declaring that an algorithm had, for the first time, passed the Turing test. The conversation had lasted a mere five minutes, and virtually no one in the field of artificial intelligence took the claim seriously.

Kurzweil nonetheless believes that a much more robust version of the test would indeed be a powerful indicator of true machine intelligence. In 2002, Kurzweil entered into a formal $20,000 bet with the software entrepreneur Mitch Kapor. The bet specifies a complex set of rules that include a three-judge panel and four contestants: the AI-powered chatbot along with three human foils.[32] The bet will be decided in Kurzweil's favor only if, by the end of the year 2029, a majority of the judges believe the AI system to be human after engaging in a two-hour, one-on-one conversation with each of the contestants. It does seem to me that passing such a test would be a strong indication that human-level AI has arrived.

Despite his distinguished career as an inventor, Kurzweil now tends to be perceived primarily as a futurist with a reasonably well-formulated theory about long-term technological acceleration, but also some seemingly outlandish—some might even say kooky—ideas about where all this progress is, in his view, likely to lead. By one account, Kurzweil takes one hundred or more supplement pills each day in the hope of prolonging his life.[33] Indeed, he believes he has already achieved "longevity escape velocity"—or in other words, he expects to repeatedly live long enough to take advantage of the next life-prolonging medical innovation.[34] Do this indefinitely, while avoiding a run-in with the proverbial bus, and you will have achieved immortality. Kurzweil told me that within about ten years, such a plan should be accessible to the rest of us. He sees the application of advanced artificial intelligence to a high-fidelity simulation of biochemistry as a critical driver of this progress. "If we could simulate biology, and it's not impossible, then we could do clinical trials in hours rather than years, and we could generate our own data just like we're doing with self-driving cars or board games or math,"[35] he told me.

Ideas like these, and perhaps in particular his earnest faith in the likelihood of his own immortality, leave Kurzweil open to a fair amount of ridicule, and many other AI researchers have a dismissive view of his hierarchical scheme for achieving general intelligence. However, one of my main takeaways from my conversation with Kurzweil was that, when it comes to the work he is doing on AI at Google, he seems extraordinary well grounded. Since he joined the company in 2012, he has been leading a team focused on merging his hierarchical theory of the brain with the latest advances in deep learning in order to produce systems with advanced language capability. One early result of his effort is the "Smart Reply" feature that can give ready answers in Gmail. While this is admittedly a far cry from human-level AI, Kurzweil remains confident in his strategy, telling me that "humans use this hierarchical approach" and that ultimately it will be "sufficient for AGI."[36]

Yet another path toward artificial general intelligence is being forged by OpenAI, a San Francisco–based research organization that was founded in 2015 with financial backing from, among others, Elon Musk, Peter Thiel and Linked-in co-founder Reid Hoffman. OpenAI was initially set up as a nonprofit entity with a mission to undertake a safe and ethical quest for AGI. The organization was conceived partly in response to Elon Musk's deep concern about the potential for superhuman machine intelligence to someday pose a genuine threat to humanity. From the onset, OpenAI has attracted some of the field's top researchers, including Ilya Sutskever, who was part of the team from Geoff Hinton's University of Toronto Lab that built the neural network that triumphed at the 2012 ImageNet competition.

In 2019, Sam Altman, who was then in charge of Silicon Valley's highest profile startup incubator, Y-Combinator, became CEO and undertook a complicated legal reshuffling that resulted in a for-profit company attached to the original non-profit entity. This was done in order to attract enough investment from the private sector so that OpenAI could fund massive investment in computational resources and compete for increasingly scarce AI talent. The maneuver paid quick dividends: in July 2019, Microsoft announced it would make a billion-dollar investment in the new company.

In the race for AGI, OpenAI is probably the best funded competitor to Google's DeepMind, although in terms of staffing levels, it remains far smaller than the more established company. Like DeepMind, OpenAI has developed powerful deep neural networks trained using reinforcement learning techniques, and its team of researchers has created systems capable of defeating the best human players at video games, such as *Dot 2*. OpenAI, however, sets itself apart with a singular focus on building larger and larger deep neural networks running on ever more powerful computational platforms. Even as others in the field warn that scalability is becoming unsustainable as a strategy, OpenAI remains deeply invested in the approach. Indeed, Microsoft's billion-dollar investment will be delivered largely in the form of compute power provided by the tech giant's cloud computing business, Azure.

OpenAI's "bigger is better" mentality has, to be sure, produced significant progress. One of the organization's most notable, and controversial, breakthroughs came with the demonstration of a powerful natural language system called GPT-2 in February 2019. GPT-2 consists of a "generative" neural network that has been trained on a massive trove of text downloaded from the internet. In a generative system, the output of a deep

neural network is essentially flipped, so that rather than identifying or classifying data—as in coming up with captions for
photographs—the system instead creates entirely new examples
that are broadly similar to the data it was trained on. Generative deep learning systems are the technology behind so-called
deepfakes—media fabrications that can be very difficult, or perhaps impossible, to distinguish from the real thing. Deepfakes
are a critical risk factor associated with artificial intelligence,
and we will discuss their implications in Chapter 8.

GPT-2 was set up so that, given a text prompt of perhaps
a sentence or two, the system would then generate a complete
narrative—in effect picking up where the prompt leaves off and
completing the story. GPT-2 caused a stir among AI researchers,
and especially the media, because the narrative text it rendered
was, in most cases, remarkably coherent. So coherent, in fact,
that OpenAI declined to share the GPT-2 code with other AI
researchers, as it had done with its previous projects, out of concern that the system could be misused. OpenAI's management
team worried, for example, that a torrent of text generated by a
GPT-2 system might clog the internet with fake news or perhaps
overwhelm e-commerce sites with fabricated product reviews.
This decision resulted in criticism from the AI research community, which values an open exchange of ideas. Then, after less
than a year, OpenAI reversed course and released the GPT-2
code after all—with little fanfare and no evidence of any malicious use. This, in turn, led to accusations that the company was
intentionally hyping its research and using alarmism to generate
publicity.

Though the prose generated by GPT-2 is generally competent in terms of grammar and structure, even a cursory reading
makes it clear that there are some real issues with the narrative
being told. Here is a portion of an example of a GPT-2 narrative
published on the OpenAI website:

Given the human-written prompt:

In a shocking finding, a scientist discovered a herd of unicorns living in a remote, previously unexplored valley, in the Andes Mountains. Even more surprising to the researchers was the fact that the unicorns spoke perfect English.

GPT-2 generated:

The scientist named the population, after their distinctive horn, Ovid's Unicorn. These four-horned, silver-white unicorns were previously unknown to science.

Now, after almost two centuries, the mystery of what sparked this odd phenomenon is finally solved.

Dr. Jorge Pérez, an evolutionary biologist from the University of La Paz, and several companions, were exploring the Andes Mountains when they found a small valley, with no other animals or humans. Pérez noticed that the valley had what appeared to be a natural fountain, surrounded by two peaks of rock and silver snow.

Pérez and the others then ventured further into the valley. "By the time we reached the top of one peak, the water looked blue, with some crystals on top," said Pérez.

Pérez and his friends were astonished to see the unicorn herd. These creatures could be seen from the air without having to move too much to see them—they were so close they could touch their horns.

While examining these bizarre creatures the scientists discovered that the creatures also spoke some fairly regular English. Pérez stated, "We can see, for example, that they have a common 'language,' something like a dialect or dialectic."[37]

The story continues for several more paragraphs, but right at the top we learn that a new species of "four-horned" unicorns has been discovered. Later, we're told that the unicorns speak "regular English," but that they "have a common 'language,' something like a dialect or dialectic." And one is left wondering what exactly to make of the sentence, "These creatures could be seen from the air without having to move too much to see them—they were so close they could touch their horns."

All this makes it pretty clear that while something is indeed coalescing within the millions of artificial neurons that make up the massive system developed by OpenAI, there is *no real understanding*. The system does not know what a unicorn is, or that a "four-horned" variety would contradict that meaning. GPT-2 suffers from the same fundamental limitation that David Ferrucci's team at Elemental Cognition and Ray Kurzweil at Google are trying to address.

In May 2020, OpenAI released GPT-3, a vastly more powerful system. While GPT-2's neural network included about 1.5 billion weights that were optimized as the network was trained, GPT-3 increased that number more than a hundredfold to 175 billion. GPT-3's neural network was trained on about half a terabyte of text, an amount so vast that the entire English version of Wikipedia—roughly six million articles—constitutes only about 0.6 percent of the total. OpenAI offered early access to a select group of AI researchers and journalists and announced plans to eventually turn the new system into its first commercial product.

Over the next few weeks, as people began to experiment with GPT-3, social media exploded with astonishment at the power of the new system. Given the proper prompts, GPT-3 could write convincing articles or poems in the style of long-dead authors. It could even generate faux conversations between historical or fictional figures. A college student used the system to generate all the posts for a self-help blog that rose to the top of the charts.[38] All this quickly led to speculation that the system represented a critical breakthrough on the path to human-level machine intelligence.

It soon became clear, however, that many of the most impressive examples had been cherry-picked from multiple trials, and that GPT-3, like its predecessor, often produced coherently written nonsense. Both of OpenAI's GPT systems are at their core powerful prediction engines. Given a sequence of words, they excel at predicting what the next word should be. GPT-3 takes this capability to an unprecedented level, and because the massive trove of text the system was trained on encapsulates real knowledge, the system does often produce very useful output. There is no consistency, however, and GPT-3 often generates rubbish and struggles with tasks that would be simple for any human.[39] Compared to its predecessor, GPT-3 can certainly write a far more compelling story about unicorns. It still, however, has no understanding of what a unicorn is.

If OpenAI simply continues to throw more computational resources at the problem, if they build even more massive neural networks, is it likely that true understanding will emerge? To me, that seems very unlikely, and many AI experts are very critical of OpenAI's continued faith in scalability. Stuart Russell, a professor of computer science at the University of California, Berkley, and the co-author of the world's leading university textbook on artificial intelligence, told me that achieving AGI will require breakthroughs that "have nothing to do with bigger datasets or faster machines."[40]

Still, the OpenAI team remains confident. In a speech at a technology conference in 2018, the company's chief scientist, Ilya Sutskever, said, "We have reviewed progress in the field over the past six years. Our conclusion is near term AGI should be taken as a serious possibility."[41] Several months later, at another conference, OpenAI CEO Sam Altman said, "I do think that much of the secret to building [AGI] is just going to be scaling these systems bigger and bigger and bigger."[42] The jury remains out on this approach, but my guess is that in order to achieve success, OpenAI will need to scale up its efforts at genuine innovation—rather than just the size of its neural networks.

A REVIVAL OF SYMBOLIC AI
AND THE DEBATE OVER INNATE STRUCTURE

As researchers wrestle with the challenges ahead, the ideas advocated by the symbolic AI camp are undergoing a kind of restoration. Nearly everyone acknowledges that the problems the symbolists tried, but largely failed, to solve must be addressed if artificial intelligence is to move forward. With the exception of a relatively small number of deep learning purists, many of whom seem to be associated with OpenAI, there is little confidence among researchers that simply scaling existing neural algorithms to take advantage of faster hardware and more data will be sufficient to produce the kind logical reasoning and common sense understanding that are essential for more general intelligence.

The good news is that this time, rather than competition between the symbolic and connectionist philosophies, we may see a reconciliation and an effort at integration. This emerging field of research has been dubbed "neuro-symbolic AI" and may represent one of the most important initiatives for the future of artificial intelligence. As decades of sometimes acrimonious competition fade into history, a new generation of AI researchers seems to be willing to try bridging the gap between the approaches. David Cox, the director of the MIT-IBM Watson AI Lab in Cambridge, MA, says that younger researchers "just don't have any of that history" and "are happy to explore intersections. They just want to do something cool with AI."[43]

There are two general schools of thought on how this integration might be accomplished. The most straightforward way may be to simply build hybrid systems that combine neural networks with software modules built using traditional programming techniques. Algorithms capable of handling logical and symbolic reasoning would somehow be linked to deep neural networks with a focus on learning. This is the strategy being pursued by David Ferrucci's team at Elemental Cognition. The second approach would be to find a way to implement symbolic AI capabilities directly into the architecture of neural networks. This might be achieved by engineering the necessary structure into deep neural networks or, I think much more speculatively, by designing both a deep learning system and a training methodology so effective that the necessary structure would somehow emerge organically. While younger researchers may be willing to consider all possibilities, among those with more established careers there continues to be a sharp debate over the best way forward.

One of the most outspoken advocates of a hybrid approach has been Gary Marcus, who until recently was a professor of psychology and neuroscience at New York University. Marcus has been a harsh critic of what he feels is an overemphasis on deep learning, and has penned essays and engaged in debates in which he argues that deep neural networks are destined to remain shallow and brittle, and that more general intelligence is very unlikely to emerge unless ideas drawn from symbolic AI are directly injected into the mix. Marcus spent much of his research career studying how children learn and acquire language and sees very little potential for a pure deep learning approach to come close to matching the remarkable capabilities of a human child. His criticisms have not always been well received by the deep learning community, where despite co-founding a machine learning startup company that was acquired by Uber in 2015, he is viewed as an outsider and someone who has not made significant contributions to the field.

Generally, experienced researchers with the most invested in deep learning tend to be dismissive of the hybrid approach. Yoshua Bengio told me that the goal should be to "solve some of the same problems that classical AI was trying to solve but using the building blocks coming from deep learning."[44] Geoff Hinton is even more disparaging of the idea, saying he doesn't "believe hybrids are the answer" and comparing such a system to a Rube Goldberg–like hybrid car in which the electric motor is used to inject gasoline into an internal combustion engine.[45] The problem is that there is, so far, no clear strategy for incorporating symbolic AI capabilities into a system built entirely from neural networks. As Marcus points out, many of deep learning's most prominent accomplishments, including Deep Mind's AlphaGo system, are in fact hybrid systems because they succeeded only by relying on traditional search algorithms in addition to deep neural networks.

As researchers argue over the efficacy of hybrid models, a parallel debate is focused on the importance of the innate structure built into machine learning systems. Though deep neural networks do often incorporate some degree of pre-designed structure—the convolutional architectures used for image recognition are one example—many hard-core deep learning advocates believe this should be kept to a minimum and that the technology will be capable of advancing from something fairly close to a blank slate. Yann LeCun, for example, told me that "in the long run we won't need precise specific structures" and points out that there is no evidence of such neural structure in the human brain, noting that "the microstructure of the cortex seems to be very, very uniform all over, whether you're looking at the visual or prefrontal cortex."[46] Researchers in this camp generally argue that innovation should be focused on developing improved training techniques that boost the ability of relatively generic neural networks to achieve greater understanding.

Researchers like Marcus, with a background in the study of cognitive development in children, push back aggressively against the "blank slate" philosophy. The brains of young children clearly include built-in capability that helps to jump-start further learning. Within days of birth, babies are able to recognize human faces. Elsewhere in the animal world, the presence of actionable intelligence that does not rely on learning is even more obvious. Anthony Zador, a neuroscientist at Cold Spring Harbor Laboratory, points out that "a squirrel can jump from tree to tree within months of birth, a colt can walk within hours, and spiders are born ready to hunt."[47] Gary Marcus often references the Alpine ibex, a species of mountain goat that lives out most of its life on steep, treacherous terrain. Newly born ibex are able to stand and navigate the slopes within hours in an environment where any kind of learning by trial and error would mean certain death. This is plug-and-play technology: it works out of the box. Researchers in this camp believe that more general, flexible artificial intelligence will likewise require built-in cognitive machinery, either injected directly into the structure of neural networks or integrated via a hybrid approach.

Deep learning advocates sometimes suggest that, though such innate structure may ultimately be important, it is likely to arise organically as part of a sustained learning process. However, if we look to the biological brain for inspiration, it does seem to me that any structure in the brain cannot be the result of long-term learning. We know that learning over an animal's lifetime does restructure the brain to some degree; it's often said that neurons that "fire together, wire together," for example. The problem is that there is no way for an individual organism to pass neural structure developed through learning during the course of its lifespan to its offspring. There is no ability to learn some stuff and then somehow cause information describing the brain structure associated with that learning to be ejected into the genetic code in the animal's egg or sperm cells. Whatever brain structure develops within an individual life dies with that organism. Therefore, it seems clear that any structure in the brain must have resulted from the normal evolutionary process, or in other words random mutations that, in rare instances, make an organism more able to thrive in its environment and are therefore more likely to be passed along to offspring. One path might be to directly copy this process through the use of evolutionary or genetic algorithms. However, directly engineering the necessary structures may be a much faster way to achieve progress.

In the debate over a hybrid versus a pure neural approach, you might say that deep learning adherents have the ultimate retort. The human brain clearly does not have some separate computer that runs special algorithms to do all the stuff that can't be handled by its neural network. It's just neurons all the way down. Still, to me it seems that the hybrid approach may be likely to produce more near-term practical results. While a purely neural implementation is clearly the path that was forged by biological evolution, this should not blind us to the possibility of faster progress using other techniques. Nor should viable approaches be dismissed simply because they are perceived as inelegant. When we landed on the moon, we didn't have a science fiction spaceship that simply zoomed down, landed and then took off again. Rather, we had a much more complicated— you might even say clunky—contraption involving a lunar module and many parts that had to be discarded along the way. Someday perhaps we'll have the science fiction spacecraft, but in the meantime we have landed on the moon.

SOME KEY CHALLENGES ON
THE PATH TO GENERAL MACHINE INTELLIGENCE

Most AI researchers recognize that significant breakthroughs will be required in order to achieve something close to human-level artificial intelligence, but there is no broad agreement on precisely what challenges are most important, or which ones should be attacked first. Yann LeCun often uses an analogy of navigating a mountain range. Only after you climb the first peak will you be able to see the obstacles that lie behind it. The hurdles that will need to be surmounted overlap and invariably intersect with the goal of building machines with the ability to truly understand natural language and engage in meaningful, unconstrained conversation. Let's look in a bit more detail at some of the key challenges that AI research will need to address. This list is not intended to be exhaustive, but a machine intelligence that cleared these hurdles would be dramatically closer to AGI than anything that exists today. Likewise, a system that was truly proficient in addressing any one of these challenges would likely spawn practical applications with enormous commercial and scientific value.

Common Sense Reasoning

What we refer to as common sense essentially amounts to a shared knowledge of the world and the way it works. We rely on common sense in nearly every aspect of our lives, but it is especially important to the way we communicate. Common sense fills in the unspoken gaps and allows us to dramatically condense our language by leaving out massive amounts of supporting information.

While virtually any adult is able to effortlessly draw on this built-in body of knowledge, doing the same has proven to be an enormous challenge for machines. Imbuing artificial intelligence with common sense is an objective that is deeply intertwined with the debates over symbolic AI versus a pure neural approach, as well as the need for structure and knowledge to be engineered into AI systems.

Recent years have seen important progress in AI systems that can analyze text and then correctly answer questions about the material. In January 2018, for example, software created by a collaboration between Microsoft and the Chinese tech giant Alibaba was able to slightly outperform the human average on a reading comprehension test created by researchers at Stanford University.[48] The Stanford test presents questions based on Wikipedia articles, in which the correct answer consists of a span of text drawn directly from the article "read" by the AI system. In other words, what we're seeing is a demonstration not of true comprehension, but rather of information extraction and pattern recognition—something that, as we've seen, deep learning systems are extraordinarily good at. When the questions require any degree of common sense reasoning or reliance on implicit knowledge of the world, performance on such tests falls off dramatically.

One of the best ways to understand the struggle that AI systems have with common sense is to look at specially formatted sentences known as Winograd schemas. Developed by Terry Winograd, a computer science professor at Stanford, these sentences leverage the power of ambiguous pronouns to test a machine intelligence's ability to employ common sense reasoning.

Here's an example:[49]

The city council refused the demonstrators a permit because they feared violence.

Who feared violence? The answer is easy for virtually anyone: the city council.

But now change just one word in the sentence:

The city council refused the demonstrators a permit because they advocated violence.

Who advocated violence?

Changing "feared" to "advocated" completely shifts the meaning of the pronoun "they." There is no way to answer this question correctly simply by extracting information from the sentence. You have to understand something about the world, specifically that a city council would prefer peaceful streets, while angry demonstrators might be inclined to violence.

Here are some other examples with the alternate words that shift the meaning of the sentence shown in brackets:

The trophy doesn't fit into the brown suitcase because it's too [small/large].

What is too [small/large]?

The delivery truck zoomed by the school bus because it was going so [fast/slow].

What was going so [fast/slow]?

Tom threw his schoolbag down to Ray after he reached the [top/bottom] of the stairs.

Who reached the [top/bottom] of the stairs?

For a series of questions like these, any normally functioning, literate adult would likely achieve a score very close to perfection. Therefore, the threshold for a passing grade should be set very high. Faced with a list of Winograd schemas, however, the best computer algorithms perform only marginally better than random guessing.

One of the most important initiatives geared toward building common sense into machine intelligence is occurring at the Allen Institute for AI in Seattle, Washington. Oren Etzioni, the CEO of the Allen Institute, told me that this effort, dubbed Project Mosaic, grew in part out of the Institute's quest to fulfill Microsoft co-founder Paul Allen's vision for an AI system that could read a chapter in a science textbook and then answer the questions at the end of the chapter. Etzioni told me that while his team's attempts to accomplish this were "state of the art," the results were less than stellar, typically resulting in a grade of around a D. One of the major stumbling blocks was the ability to handle common sense and logical reasoning while answering the questions. It's fairly easy for an AI system to learn factual material about photosynthesis from a biology textbook, for example. But the real challenge, says Etzioni, is when you have a question like "If you have a plant in a dark room and you move it nearer the window, will the plant's leaves grow faster, slower or at the same rate?"[50] This requires understanding that there will be more light closer to the window and the ability to reason that this will allow the plant to grow faster.

The first goal of Project Mosaic is to create a standard set of benchmarks designed to measure the ability of machines to exhibit common sense. Once this is complete, the Institute plans to deploy a variety of techniques, including "crowdsourcing, natural language processing, machine learning, and machine vision"[51] to generate the built-in knowledge of the world that will be required to imbue an AI system with common sense.

While Etzioni and his team are strong believers in using a hybrid approach that will bring together a variety of techniques, this idea, as you might expect, generates little enthusiasm among the staunchest deep learning advocates. When I asked Yoshua Bengio if he thought efforts like Project Mosaic were important, or if he thought common sense reasoning might somehow emerge organically from the learning process, he left no doubt about his faith in a deep learning approach: "I'm sure common sense will emerge as part of the learning process. It won't come up because somebody sticks little bits of knowledge into your head, that's not how it works for humans."[52] Yann LeCun likewise believes the path to common sense is through learning, telling me that Facebook's AI research team is working on "getting machines to learn by observation from different data sources—learning how the world works. We're building a model of the world so that perhaps some form of common sense will emerge and perhaps that model could be used as kind of a predictive model that would allow a machine to learn the way people do."[53]

The good news is that both approaches are being pursued aggressively by some of the world's brightest AI researchers. A breakthrough that results in an AI system that is able to reliably deploy the kind of common sense reasoning that we take for granted in humans would be an extraordinary advance, regardless of whether it emerges organically or results from a more engineered approach.

Unsupervised Learning

As we've seen, the two primary techniques used to train deep learning systems are supervised learning, which requires large quantities of labeled data, and reinforcement learning, which requires a huge number of trial-and-error iterations as an algorithm attempts to succeed at a task. Though human beings likewise employ these techniques, they constitute only a tiny fraction of the learning that goes on in a young child's mind. Very young children learn from simple observation, by listening to the voices of their parents and by engaging and experimenting directly with the world around them.

Newly born babies begin the process almost immediately, learning directly from their environment long before they have the physical ability to interact with it in any deliberate way. Somehow, they manage to develop a physical model of the world and begin to build the base of knowledge that underlies common sense. This ability to learn directly and without assistance from structured and labeled data is known as "unsupervised learning." This remarkable ability may well be enabled by some kind of cognitive structure built into the child's brain, but there is no doubt that the ability of a human child to learn independently, and especially to acquire language, vastly outpaces anything that can be accomplished with the most powerful deep learning system.

This early unsupervised learning then supports more advanced knowledge acquisition later on. Even when an older child's learning is to some extent supervised, the training data required is a tiny fraction of what would need to be provided to even the most advanced algorithm. A deep neural network might require many thousands of labeled training photographs before it can reliably attach the names of animals to their images. In contrast, a parent pointing to an animal and saying, "This is a dog" a single time might well be sufficient. And once the child can identify the animal, she can likely do so in any configuration; the dog could be sitting or standing or running across the road, and still the child can consistently attach a name to it.

Unsupervised learning is currently one of the hottest research topics in the field of artificial intelligence. Google, Facebook and DeepMind all have teams focused in this area. Progress, however, has been slow, and few if any truly practical applications have so far emerged. The truth is that no one really has any idea exactly how the human brain achieves its unparalleled competence at autonomously learning from unstructured data. Most current research is focused on less ambitious-sounding variants of unsupervised learning, such as predictive learning or self-supervised learning. Example projects might include trying to predict the next word in a sentence or the image that makes up the next frame in a video. While these kinds of tasks may seem a far cry from what humans manage, many researchers believe that the ability to make predictions is absolutely central to intelligence and that experiments like these will drive things in the right direction. It's difficult to overstate the magnitude of a genuine breakthrough in unsupervised machine learning. Yann LeCun, for example, believes it may well be the gateway that leads to progress on nearly every other aspect of general intelligence, saying, "Until we figure out how to do this . . . we're not going to make significant progress because I think that's the key to learning enough background knowledge about the world so that common sense will emerge. That's the main hurdle."[54]

Understanding Causation

Students studying statistics are often reminded that "correlation does not equal causation." For artificial intelligence, and especially deep learning systems, understanding ends at correlation. Judea Pearl, a renowned computer scientist at UCLA, has over the past thirty years revolutionized the study of causation and constructed a formal scientific language for expressing causal relationships. Pearl, who won the Turing Award in 2011, likes to point out that while any human understands intuitively that the sunrise causes a rooster to crow, rather than vice versa, the most powerful deep neural network would likely fail to achieve a similar insight. Causation cannot be derived simply by analyzing data.[55]

Human beings have a unique ability to not just detect correlations but also to understand causal effects, and we can do so on the basis of remarkably few examples. Joshua Tenenbaum, a professor of computational cognitive science at MIT, who describes his research focus as "reverse engineering the human mind" in the hope of gaining insights that will be useful in building smarter AI systems, points out:

> Even young children can often infer a new causal relation from just one or a few examples—they don't even need to see enough data to detect a statistically significant correlation. Think about the first time you saw a smartphone, whether it was an iPhone or some other device with a touchscreen where somebody swipes their finger across a little glass panel, and suddenly something lights up or moves. You had never seen anything like that before, but you only need to see that once or a couple of times to understand that there's this new causal relation, and then that's just your first step into learning how to control it and to get all sorts of useful things done.[56]

Understanding causal relationships is critical to imagination and to the generation of mental counterfactual scenarios that enable us to solve problems. Unlike a reinforcement learning algorithm that needs to fail thousands of times before figuring out how to succeed, we can run a kind of simulation in our heads and explore the likely outcomes of alternate courses of action. This would be impossible without an intuitive grasp of causation.

Researchers like Pearl and Tenenbaum believe that an understanding of causal relationships—in essence the ability to ask and answer the question "Why?"—will be an essential ingredient in building more general machine intelligence. Pearl's work on causation has had an enormous impact in the natural and social sciences, but he believes that AI researchers have largely failed to get the memo and have generally been too focused on the correlations that are so efficiently identified by machine learning systems.[57] This is changing, however. For example, Yoshua Bengio and his team at the University of Montreal recently published important research on an innovative way to build an understanding of causation into deep learning systems.[58]

Transfer Learning

Graham Allison, a political scientist and professor at Harvard, is known for coining the phrase "Thucydides's Trap." The term references the Greek historian Thucydides's *History of the Peloponnesian War*, which chronicles the conflict between Sparta and a newly ascendant Athens in the fifth century BC. Graham believes that the war between Sparta and Athens represents a kind of historical principle that remains applicable today. In his 2017 book *Destined for War*, he argues that the United States and China are caught in a contemporary Thucydides's Trap and that as China continues to rise in power and influence, conflict may well be inevitable.[59]

Could an artificial intelligence system read a historical document like the *History of the Peloponnesian War* and then successfully apply what it learns to a contemporary geopolitical situation? To do so would be to reach one of the most important milestones on the path toward artificial general intelligence: transfer learning. The ability to learn information in one domain and then successfully leverage it in other domains is one of the hallmarks of human intelligence and is essential to creativity and innovation. If more general machine intelligence is to be genuinely useful, it will have to do more than simply answer the questions at the end of the chapter; it will need to be able to apply what it learns, and any insights it develops, to entirely new challenges. Before an AI system has any hope of accomplishing this, it will need to move well beyond the superficial level of understanding that currently coalesces within deep neural networks and achieve genuine comprehension. Indeed, the ability to apply knowledge in a variety of domains and in novel situations may turn out to be the single best test for true understanding in a machine intelligence.

THE PATH TO HUMAN-LEVEL ARTIFICIAL INTELLIGENCE

Nearly all the AI researchers I have spoken to believe that human-level artificial intelligence is achievable and someday will be inevitable. To me, this seems reasonable. The human brain, after all, is fundamentally a biological machine. There is no reason to believe that there is anything magical about biological intelligence or that something broadly comparable couldn't someday be instantiated in an entirely different medium.

Indeed, a silicon-based substrate would seem to have many advantages over the biological wetware that powers the human brain. Electronic signals propagate at vastly higher speeds in computer chips than in the brain, and any machine that someday equaled our ability to reason and communicate would continue to enjoy all the advantages that computers currently have over us. A machine intelligence would benefit from flawless memory, even of events that occurred deep in the past, and would have the ability to calculate and to sift and search through enormous troves of data at fantastic speed. It would also be able to directly connect to the internet or to other networks and tap into virtually limitless resources; it would effortlessly talk to other machines, even as it mastered conversation with us. In other words, human-level AI, from it very inception, would in a great many ways be superior to us.

Though faith in someday reaching this destination is nearly universal, the route that will take us there, and the time of arrival, remain shrouded in deep uncertainty. So far, progress has largely been incremental. For example, in late 2017, DeepMind released AlphaZero, an update to its Go-playing AlphaGo system. AlphaZero dispensed with the need for a supervised learning regimen on data from thousands of Go matches played by humans and instead began with essentially a blank slate, learning to play at superhuman levels purely on the basis of simulated games played against itself. The system also had the ability to be trained for other challenges, including chess and the Japanese game shogi. AlphaZero quickly demonstrated that it is the top chess-playing entity on the planet by defeating the very best dedicated chess-playing algorithms—which, of course, were already able to easily dispense with the most capable human players. Demis Hassabis told me that AlphaZero probably represents a general solution to "information complete" games, or in other words the type of challenges in which all the information you need to prevail is readily available as game pieces on a board or pixels on a screen.

The real world in which we live is, of course, far from information complete. Nearly all the most important areas in which we would someday like to leverage advanced artificial intelligence require the ability to operate under uncertainty and to deal with situations where vast amounts of information are hidden or simply unattainable. In January 2019, DeepMind again demonstrated progress with its release of AlphaStar, a system designed to play the strategy video game *StarCraft*. *StarCraft*

simulates a galactic struggle for resources between three different extraterrestrial species, each of which is controlled in real
time by an online player. *StarCraft* is not an information complete game; rather, players need to "scout" in order to discover
hidden information about their opponents' activities. The game
also requires long-term planning and management of resources
across a vast game space. In another first for DeepMind's team,
AlphaStar defeated a top professional *StarCraft* player 5-0 in a
match conducted in December 2018.[60]

Though these achievements are impressive, they still do not
come close to overcoming the major limitations that confine today's AI systems to highly specific, narrow domains. AlphaStar,
for example, must be extensively trained, using supervised and
reinforcement learning techniques, to play in the role of a particular alien species. Switching to a different species, with different
relative strengths, requires re-training from scratch. Likewise,
AlphaZero can easily achieve world-dominating chess or shogi
capability, but the system would not be able to beat a child at
a game of checkers without retraining on that game. Even the
most powerful systems at the very forefront of AI research remain shallow and brittle. And as the Allen Institute's Oren Etzioni likes to point out, any of these systems would continue to
play unperturbed if they learned the room was on fire.[61] There
is no common sense, no true understanding.

How long might it take to overcome these limitations and
succeed in building a genuine thinking machine? As I engaged
in the conversations that are recorded in my book *Architects of
Intelligence*, I conducted an informal survey of the top minds
in the field of AI. I asked each of the twenty-three individuals I
spoke with to give me his or her prediction for the year artificial
general intelligence would have at least a fifty percent probability of being achieved. Most of the participants requested that

their guesses remain anonymous. Five of the researchers I spoke with declined to make a prediction at all, pointing out that the path to human-level AI is highly uncertain and that an unknown number of specific challenges will need to be overcome. Nonetheless, eighteen of the world's foremost AI experts did give me their best guess, and I think the results, which are shown in the table below, are very interesting.[62]

Year AGI Achieved	Years from 2021	Number of Guesses
2029	8	1 (Kurzweil)
2036	15	1
2038	17	1
2040	19	1
2068	47	3
2080	59	1
2088	67	1
2098	77	2
2118	97	3
2168	147	2
2188	167	1
2200	179	1 (Brooks)

Note that these guesses were made in 2018, which explains the preponderance of years ending in "8." A guess of 2038, for example, was actually a guess of "twenty years from now." I strongly suspect that if I asked the same people to guess again today, I would get essentially the same estimates, moving these numbers out by three years or so. This does raise the concern that the achievement of AGI might fall prey to the old joke that physicists have often told about nuclear fusion: "It's always thirty years in the future."

The average guess was the year 2099, or roughly eighty years from now.*,63 The predictions are neatly bracketed by guesses from two people who were willing to go on the record. Ray Kurzweil, as we've seen, remains adamant that human-level AI will come into existence by 2029—now just eight years away. Rodney Brooks, a co-founder of iRobot Corporation and widely regarded as one of the world's foremost roboticists, thinks it will take nearly 180 years for AGI to arrive. This gaping chasm between predictions—with multiple researchers anticipating human-level AI within a decade or two, whereas others think it could be centuries—is, I think, a vivid illustration of just how unpredictable the future of artificial intelligence is likely to be.

THE QUEST TO build human-level AI is, I think, the single most fascinating topic in the field of artificial intelligence. Someday it may well result in humanity's most consequential and disruptive innovation. In the meantime, however, artificial intelligence as a practical tool will remain relatively narrow, and in many ways quite limited. To be sure, AI systems designed to solve real-world problems will be continuously upgraded as research at the very frontier of the field is incorporated, but for the foreseeable future, the power of this new technology will be delivered not by a single, highly flexible machine intelligence, but rather by an explosion of specific applications that are already beginning to scale across nearly every aspect of industry, the economy, society and even culture.

Without question, AI has the potential to deliver profound benefits, especially in critical areas like healthcare, scientific

---

* This average is pessimistic relative to other surveys that have been conducted. These have included a much larger number of AI researchers with widely different experience levels, often at AI conferences. Most results have clusters around the years 2040 to 2050 for AGI with a fifty percent probability. See endnote 63, Chapter 5, for a list of such surveys.

research and broad-based technical innovation. There is another side to the technology, however. Artificial intelligence will come coupled with unprecedented challenges and dangers—to jobs and the economy, to personal privacy and security and perhaps ultimately to our democratic system and even to civilization itself. These risks will be the primary focus of the next three chapters.

CHAPTER 6

DISAPPEARING JOBS AND THE ECONOMIC CONSEQUENCES OF AI

IN MY 2015 BOOK *RISE OF THE ROBOTS: TECHNOLOGY AND THE Threat of a Jobless Future*, I argued that advances in artificial intelligence and robotics would eventually destroy a great many jobs that tended to be routine and predictable—potentially leading to increased inequality and structural unemployment. As I began writing this book in January 2020, I assumed that the main task before me in this chapter would be to defend that thesis in the face of the longest economic recovery since World War II and a headline unemployment rate of about 3.6 percent.

Needless to say, the coronavirus pandemic and the ensuing shut down of economies in the United States and across the globe have led us into an entirely new economic reality. Nonetheless, I believe the arguments I planned to make before the crisis emerged remain highly relevant. Even in a time of historically low unemployment, I believe that the trends I discussed in *Rise of the Robots* remained firmly in play, and that the relative prosperity suggested by economic indicators in the years leading up to the current crisis was, at least to some extent, an illusion. In the wake of the pandemic, the trend toward increased job automation may well be amplified and could have a dramatic impact as we look forward to recovery from the current economic disaster.

Imagine that you are an American economist in the year 1965. As you gazed out over the U.S. economy and job market, you would see that about ninety-seven percent of men between the ages of twenty-five and fifty-four—old enough to have completed schooling but too young to retire—are either employed or actively seeking work. This would seem entirely expected and normal to you. Now suppose that a time traveler from the future appears and tells you that in the year 2019, only about eighty-nine percent of prime working age men will be in the workforce—and that by 2050 the fraction of American men in this age group who are completely disenfranchised from the job market might well grow to a quarter or even a third.*,[1]

It seems like a good bet that you would find this alarming. Perhaps the phrase "mass unemployment" would even cross your mind. You would surely wonder just what all those non-working men were up to. But now the time traveler tells you that the headline unemployment rate reported by the government in 2019 is significantly below four percent and that interest rates are below the 1965 level. Both measures, points out the time traveler, are close to historic lows. Furthermore, you are told, the U.S. Federal Reserve, rather than planning to raise interest rates, is signaling that it might well lower them further in an effort to boost the economy.

All of these things would likely be quite surprising and confusing to an economist from the mid- to late twentieth century.

---

* Our time traveler is based on former Treasury Secretary and Director of the National Economic Council Lawrence Summers, who made the estimate of a quarter to a third of working age men being out of the workforce by 2050 in November 2016. (See endnote 1, Chapter 6.)

As we'll see in this chapter, the economy and job market in the United States, as well as in many other developed countries, are now operating in ways that seem to defy many of the rules and assumptions that once appeared to be solidly backed by empirical evidence.

In *Rise of the Robots*, I argued that these changes are being driven largely by accelerating progress in information technology. A long list of key innovations is now behind us—advancing factory automation, the personal computer revolution, the internet, the rise of cloud computing and mobile technology—and the resulting transformation has been playing out over the course of decades. The most important technological impact, however, still lies in the future. The rise of artificial intelligence has the potential to upend both the job market and our overall economic system in ways that are far more dramatic—and foundational—than anything we have seen previously.

As we stand at the leading edge of the coming disruption, there are good reasons to be concerned. The transformations that have occurred over just the past decade or two have arguably played an important role in unimaginable political upheaval and have rended the very fabric of society. Studies, for example, have shown a direct correlation between regions in the United States most vulnerable to job automation and voters who strongly supported Donald Trump in the 2016 presidential election.[2] Before the coronavirus pandemic upended our lives, there was more focus on another health crisis that has been devastating the United States, and areas that experienced substantial middle class job loss also tended to be on the front lines of the opioid epidemic.[3] If the changes we've seen so far pale in comparison with what might come, there is a real risk of future social and economic disruption on an unprecedented scale—as well as the rise of even more dangerous political demagogues who will thrive on the fear that is certain to accompany such a rapidly shifting landscape.

The reality is that artificial intelligence will be a dual-edged sword in terms of its economic impact. On one hand, it will likely increase productivity, make products and services more affordable and enable innovation that can improve all our lives. AI has the potential to create economic value that will be indispensable as we look toward digging ourselves out of the massive economic hole in which we now find ourselves. On the other hand, it is virtually certain to eliminate or deskill millions of jobs while driving economic inequality to even higher levels. Aside from the social and political implications of unemployment and ever-rising inequality, there is another important economic consequence: a vibrant market economy depends on vast numbers of consumers who are able to purchase the products and services being produced. If these consumers do not have jobs, and thus income, how will they create the demand necessary to drive continued economic growth?

AI AND JOB AUTOMATION: IS THIS TIME DIFFERENT?

The fear that machines might someday displace workers and produce long-term, structural unemployment has a long history, stretching back, at a minimum, to the Luddite revolts that took place in Nottingham, England, over two hundred years ago. In the decades since, the alarm has been raised again and again. In the 1950s and 1960s, for example, there was a great deal of concern that industrial automation would soon displace millions of factory jobs, leading to widespread unemployment. So far, however, history shows that the economy has generally adjusted to advancing technology by creating new employment opportunities and that these new jobs often require more skills and pay higher wages.

One of the most extreme historical examples of technologically-induced job losses—and a case study often cited by those who are skeptical that technological unemployment will ever pose a problem—concerns the mechanization of agriculture in the United States. In the late 1800s, about half of American workers were engaged in farming. Today, the number is between one and two percent. The advent of tractors, combine harvesters and other agricultural technology irreversibly vaporized millions of jobs. This transition did result in significant short- and medium-term unemployment as displaced farm workers migrated to cities in search of factory work. Eventually, however, the unemployed workers were absorbed by a rising manufacturing sector, and over the long run, average wages as well as overall prosperity increased dramatically. Later, factories automated or moved offshore, and workers transitioned again, this time to the service sector. Today, nearly eighty percent of the American workforce is employed in service industries.

The key question is whether the job market disruption that results from the impact of artificial intelligence will lead to a similar outcome. Is AI just another example of a labor-saving innovation like the agricultural technology that transformed farming? Or is it something fundamentally different? My argument has been that AI is indeed different, and the reason is anchored in the core thesis of this book: that artificial intelligence is a systemic, general-purpose technology not unlike electricity, and it will therefore ultimately scale across and invade every aspect of our economy and society.

Historically, labor market technological disruptions have tended to impact on a sector-by-sector basis. Agricultural mechanization destroyed millions of jobs, but a rising manufacturing sector was available to eventually absorb those workers. Likewise, as manufacturing automated and factories offshored to low-wage countries, a rapidly growing service sector provided opportunities for the displaced workers. In contrast, artificial intelligence will impact every sector of the economy more or less simultaneously. Most importantly, this will include the service sector and white collar jobs that now engage the vast majority of the U.S. workforce. AI's tentacles will eventually reach into and transform virtually every existing industry, and any new industries that arise in the future will very likely incorporate the latest AI and robotics innovations from their inception. In other words, it seems very unlikely that some entirely new sector with tens of millions of new jobs will somehow materialize to absorb all the workers displaced by automation in existing industries. Rather, future industries will be built on a foundation of digital technology, data science and artificial intelligence—and as a result, they will simply not generate large numbers of jobs.

A second point involves the nature of the activities undertaken by workers. It's reasonable to estimate that roughly half our workforce is engaged in occupations that are largely routine and predictable in nature.[4] By this, I don't mean "rote-repetitive" but simply that these workers tend to face the same basic set of tasks and challenges again and again. In other words, the essence of the job—or at least a large fraction of the tasks that comprise it—is essentially encapsulated in historical data reflecting what the worker has done over time. Such data will eventually provide a rich resource for machine learning algorithms that can be turned loose to figure out how to automate many of these tasks. In other words, we're facing a future where nearly all kinds of routine, predictable work are eventually going to evaporate, and this is likely to prove an especially difficult challenge for those workers who are best suited to such work. Throughout the twentieth century, advancing labor-saving technology drove workers to move to different sectors, but for the most part, they continued doing largely routine work. Imagine the transition from a farm worker in 1900, to a factory assembly line worker in 1950, to a cashier scanning barcodes at Walmart today. These are all very different jobs in entirely different sectors, but they are all defined by largely routine and predictable tasks. This time around, there aren't going to be large numbers of routine jobs in some new sector to accommodate displaced workers. Instead, workers will be faced with making an entirely different kind of transition into work that is fundamentally non-routine and may often require qualities such as the ability to effectively build relationships with others or to perform non-routine analytical or creative work. Assuming that a sufficient number of such new jobs are available, some workers will successfully make this transition, but many others will likely struggle.

In other words, I think we face a scenario where a significant fraction of our workforce is eventually at risk of being disenfranchised from the job market. But is there any evidence that anything like this has actually occurred? After all, the unemployment rate prior to the advent of the coronavirus pandemic was well below four percent.

THE STORY UP UNTIL
THE ONSET OF THE CORONAVIRUS PANDEMIC

Over the ten-year period from the end of the Great Recession in 2009 until January 2020, the longest postwar economic recovery on record, the unemployment rate fell from ten percent to 3.6 percent—a level lower than any recorded in the past fifty years.[5] An important caveat, however, is that this headline unemployment rate, which is measured on the basis of a household survey conducted by the U.S. Census Bureau, captures only those workers who are actively seeking employment. Anyone who would like to have a job but has become disheartened and given up or who believes there are no jobs available that they would be willing to accept, is not counted as unemployed.

To get some insight into the number of people who have become completely detached from the labor force, it's useful to look at the labor force participation rate. The story here is far less positive than the headline unemployment rate.

As Figure 1 shows, the percentage of prime age working men in work or actively seeking employment has fallen from about ninety-seven percent in 1965 to a nadir of eighty-eight percent in 2014 before recovering slightly to about eighty-nine percent in January 2020.[6] The number of men entirely disenfranchised from the job market has nearly quadrupled over this time. One destination for men who have been exiting the job market appears to be the Social Security Disability program, which saw a surge of applications between 2007 and 2010.[7] Given that there was no evidence of an epidemic of workplace injuries, it seems likely that the program is being used as an income of last resort by workers who see few viable job market opportunities. While the impact on labor force participation for men has been the most dramatic, the overall statistics show a broadly similar story over the two decades since the turn of the century.

Figure 2 shows the workforce participation rate for all workers aged eighteen to sixty-four, including both men and women.[8] The rising participation rate up to the year 2000 reflects the entry of more women into the workforce. After that peak, however, the trend has been downward as both men and women have exited the labor market. In other words, even as the unemployment rate fell to historic lows, there has been an ever-increasing mass of completely disenfranchised workers who remained largely invisible as the overall narrative pointed to a booming job market. While technological change has certainly not been the only factor at play here, the relentless automation of well-paid routine jobs in factories and offices has likely played a significant role.

A second important trend involves the decoupling of productivity and wages, along with a relentless drive toward increased

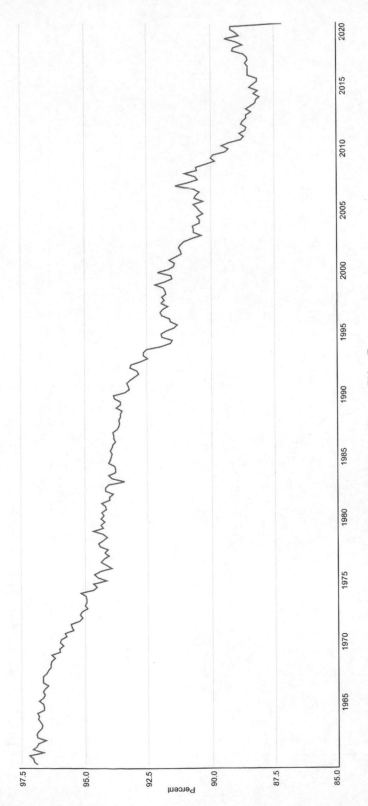

Figure 1. Workforce Participation Rate, Men Ages Twenty-Five to Fifty-Four

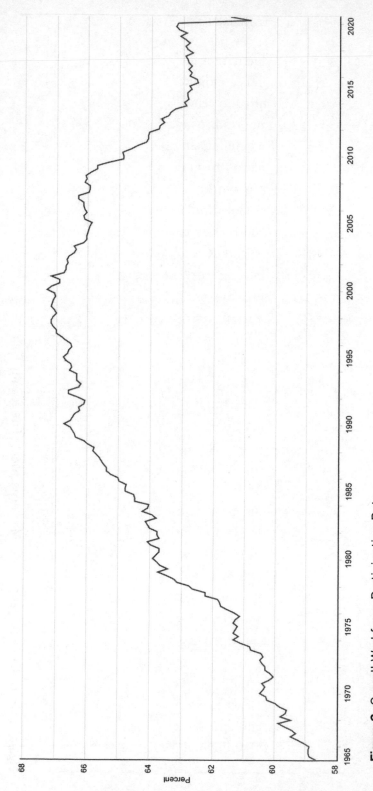

Figure 2. Overall Workforce Participation Rate

inequality. Labor productivity is a measure of worker effectiveness and is equal to total economic output divided by the number of labor hours required to generate that output. Productivity
is perhaps the most important of all economic metrics. A high
rate of productivity is a defining characteristic that distinguishes
a wealthy, developed nation from an impoverished one. As the
technology employed in workplaces advances, and as other
factors such as worker education and health likewise improve,
workers can produce more. As a result, they should be able to
command higher wages, and therefore rising productivity essentially deposits money into the pockets of nearly all workers and
is a critical driver of broad-based national prosperity. At least
that is the standard economic narrative.

As Figure 3 shows, however, at least since the 1970s, compensation for workers has failed to track rising productivity and
an ever-widening chasm has opened between the two lines.[9] The
upshot of this is that nearly all the gains from technological
progress and improving productivity are now being captured by
a relatively small group of people near the top of the income distribution. In other words, business owners, managers, superstar
employees and investors are capturing the fruits of progress,
and ordinary workers are getting almost nothing. It's worth noting that this graph reflects compensation for all workers in the
business sector, and that includes top-level executives, superstar
athletes and entertainers, as well as other highly paid workers.
If the graph instead reflected only the average non-supervisory
workers who account for about eighty percent of the U.S. workforce, the gap between productivity and compensation would be
even greater.

I would argue that the widening divergence between these
two lines is driven, at least in part, by the changing nature of
the machines and technology deployed in workplaces. During
America's "golden age" following World War II, the two lines

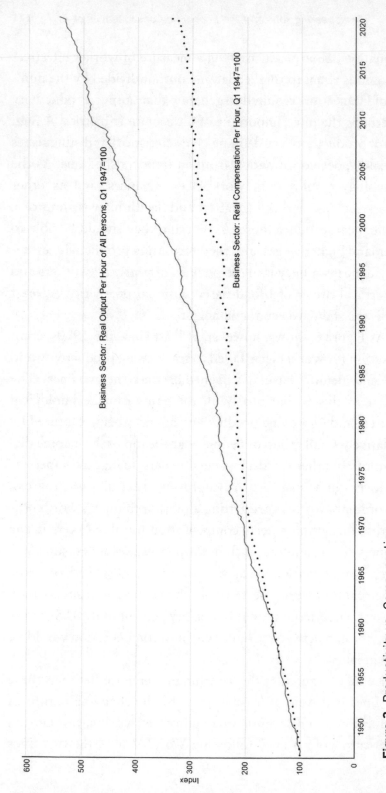

Business Sector: Real Output Per Hour of All Persons, Q1 1947=100

Business Sector: Real Compensation Per Hour, Q1 1947=100

Figure 3. Productivity vs. Compensation

on the graph were tightly coupled and the machines used in workplaces were clearly tools operated by workers; as the tools improved, the workers' output increased and they became more valuable. As technology has continued to advance in the decades since, however, many of the machines used in workplaces are gradually becoming more autonomous, and technology is increasingly substituting for, rather than complementing, labor. In other words, technology is now making an increasing fraction of workers less rather than more valuable. This, in turn, makes workers more interchangeable, reduces their bargaining power, and acts to push down compensation even as productivity continues to increase.

This decoupling of productivity and compensation leads directly to increased income inequality. As technology displaces or diminishes the value of labor, a larger share of business profits is captured by capital. This decline in labor's share of national income has been found over the last two decades in the United States as well as in a variety of other developed countries. Because capital ownership is highly concentrated in the hands of the wealthy, a redirection of income from labor to capital amounts to a redistribution from the many to the few, and this increases income inequality. In the United States the trend has been especially dramatic and is demonstrated vividly by the rise of the Gini coefficient. This index is a measure of the concentration of wealth. At the extremes, a Gini value of zero would indicate that everyone in a country has an equal share of wealth, and a value of 100 would mean that a single individual owns all the nation's wealth. Realistic values generally fall between roughly 20 and 50, with a higher number indicating more inequality. In the U.S., the Gini coefficient rose from 37.5 in 1986 to 41.4 in 2016—a level higher than any previously recorded.[10]

This trajectory toward rising income inequality has been driven in part by a general decline in the quality of jobs on offer in the United States. In recent decades, American job creation has been weighted increasingly toward low-wage jobs in the service sector. These jobs, in areas like retail sales, food preparation and serving, security and cleaning or janitorial jobs in offices and hotels, provide minimal incomes and few if any benefits, and are often less than full-time with unreliable hours. The rise of the gig economy, in which workers receive payments based on a task-completion basis with virtually no guarantee of a predictable income and little or no access to the legal safeguards provided to other workers, has further exacerbated the trend. A November 2019 report from the Brookings Institution found that a full forty-four percent of the U.S. workforce is engaged in low-wage jobs providing income averaging about $18,000 per year.[11]

This change in the nature of the jobs available to American workers was made especially evident when a group of researchers developed a new economic metric in 2019. The U.S. Private Sector Job Quality Index measures the ratio of good jobs, defined as those that provide an above-average income, to low-quality jobs, which offer an income below the average.[12] An index value of 100 indicates an equal number of good- and low-quality jobs, while a value below 100 indicates that lower-quality jobs dominate the employment landscape. Over the 30 years between 1990 and the end of 2019, the index plunged from 95 to 81.[13] This decline in quality is likely closely tied to the evaporation of largely routine but well-paying jobs in environments like factories and offices. These are the jobs that once formed the backbone of the American middle class but have been relentlessly destroyed by both technology and globalization.

The economy has, of course, also created higher-skill, higher-paying jobs, but these are rarely accessible to the nearly three quarters of American workers who lack a four-year college degree. And even among college graduates, underemployment is a serious and growing problem. Stories of college graduates weighed down by massive student loan obligations while working as baristas or fast food workers are all too common. Data published by the Federal Reserve Bank of New York in February 2020 showed that a full forty-one percent of recent college graduates are engaged in jobs that do not require a college degree. For college graduates as a whole, the underemployment figure is one in three. And recent college graduates aged twenty-two to twenty-seven had an unemployment rate of more than six percent even as headline unemployment across the economy as a whole fell to 3.6 percent.[14] In other words, even as the conventional wisdom suggests that we need to put more emphasis on education and expand college enrollment, the economy is simply not creating enough skilled job opportunities to absorb the graduates already being produced.

The rise in income inequality and the decline in job quality are not just bad news for the individuals directly impacted. Rather, they undermine the market demand required to drive us toward sustained economic vitality. Roughly seventy percent of the U.S. economy is associated directly with individual consumer spending. Even that fraction, however, underestimates the importance of consumer demand because business investment is also tied to consumer demand. For example, think about how airplanes produced by Boeing—certainly not a consumer item—are purchased by airlines only if they, in turn, anticipate consumer demand for plane tickets. This economic dependency has, of course, been brought into stark relief by the impact of the coronavirus crisis.

Jobs are the primary mechanism that delivers purchasing power into the hands of consumers. As the distribution of income becomes more unequal, the bulk of workers, and therefore consumers, are left with less discretionary income. Over the past few decades, income has risen dramatically for the wealthy few, but this small fraction of the population simply cannot and will not spend to a degree that makes up for the loss of discretionary income at lower levels of the income distribution. In other words, the broad-based consumer demand for products and services that is vital to generating economic growth is gradually eroding.

Evidence for underwhelming consumer demand has manifested in the breakdown in the normal relationship between unemployment and inflation. In 1958, the economist William Phillips showed that there is generally a consistent trade-off between unemployment and inflation. As unemployment falls, inflation rises. When I studied economics in college, this inverse relationship, known as the Phillips curve, was taught as one of the basic principles of the field. In the years since the end of the Great Recession in 2009, however, this relationship has broken down and low unemployment now coexists with very low rates of inflation and low interest rates.[15] I believe that an important reason for this is that falling unemployment is no longer associated with increases in wages or consumer demand that are sufficient to drive inflation. As advancing technology, along with globalization, has eroded the ability of most average workers to bargain for higher wages, the mechanism that gets purchasing power into the hands of consumers and drives increasing demand has become less and less effective.

More evidence comes from the fact that large U.S. corporations have been sitting on enormous amounts of cash, much of which is invested in U.S. Treasury securities paying historically low interest rates. As of the end of 2018, American businesses were hording about $2.7 trillion.[16] If the executives running these companies saw evidence of vibrant demand for goods and services, why wouldn't they invest more of this money in developing new products or ramping up production to meet that increasing demand? Without robust demand, the U.S. economy has managed only middling rates of growth and has become dependent on the Federal Reserve holding interest rates at un-usually low levels even as the unemployment rate fell below four percent.

Another important implication of tepid consumer demand is that it undermines productivity growth. Economists who are skeptical of the impact of artificial intelligence and robotics on the job market are quick to point out that if machines were indeed substituting for labor at a rapid clip, we should see soaring labor productivity as the remaining workers produced ever more output. In the absence of skyrocketing productivity, the economists brush aside any concerns about robots stealing jobs. The problem with this assertion is that output is entirely dependent on demand. No business will continue to produce goods or services unless there is a customer prepared to purchase that output. (I also explained the idea that productivity is limited by demand in more detail in my 2015 book *Rise of the Robots*. I find it somewhat surprising that economists don't focus more on this issue and instead tend to simply declare that a lack of "soaring productivity" proves that job automation is a non-issue.[17])

Imagine a worker whose job is to give haircuts. The productivity of that worker might be measured in terms of the number of haircuts performed per hour. Lots of things would affect that rate of productivity. Does the worker have good training and quality tools? Is there a stable supply of electricity to keep the equipment running? These are the kinds of things that economists tend to focus on. But there is something else that is absolutely critical: the number of customers who show up for a haircut. If there is a long line of eager customers, productivity will be high. If only an occasional client wanders in, productivity will be low—regardless of how good the training or haircutting technology is.

This idea that productivity growth is limited by demand came up when I spoke with James Manyika, the chairman of the McKinsey Global Institute (MGI), which has conducted numerous important studies focused on the impact of technology on businesses and the economy. As Manyika explained:

> We also know the critical role of demand—most economists, including here at MGI, have often looked at the supply-side effects of productivity, and not as much at the demand side. We know that when you've got a huge slowdown in demand you can be as efficient as you want in production, and measured productivity still won't be great. That's because the productivity measurement has a numerator and a denominator: the numerator involves growth in value-added output, which requires that output is being soaked up by demand. So, if demand is lagging for whatever reason, that hurts growth in output, which brings down productivity growth, regardless of what technological advances there may have been.[18]

The bottom line is that the years leading up to the onset of the coronavirus pandemic delivered an American economy that was a bit like a shiny, newly painted car—but with serious problems under the hood. The unemployment rate looked great, but a large and growing fraction of the population was being left completely behind. Inequality has risen dramatically, and most workers are no longer experiencing increased prosperity as a result of technological progress. And as things have become ever more unequal, the mechanism that distributes the income that powers consumer demand is eroding, and that in turn is undermining economic growth and dampening down the sustained rise in productivity that is crucial to future prosperity. The pandemic has completely upended things and plunged us into an unprecedented economic crisis, but all these trends remain in place and will likely produce headwinds that will make it even more challenging to recover from our current predicament.

POST COVID-19 AND RECOVERY

The coronavirus pandemic has unleashed a global economic crisis of unprecedented ferocity. In the United States and in countries throughout the world, millions of jobs have been lost very nearly overnight, entire sectors have been virtually shut down and the economy has plunged into the deepest downturn since the Great Depression in the 1930s. As of December 2020, the unemployment rate stands at nearly seven percent, and all indications are that things could well get worse before wide deployment of vaccines begins to bend the curve sometime around the middle of 2021. The U.S.'s bungled management of the pandemic led to a widespread resurgence of the virus, and the country is recording more than 4,000 COVID-19 deaths in a single day as of January 2021. As hospitalizations surge, U.S. state and local governments are once again forcing businesses to close, while in the United Kingdom and many European countries national lockdowns are again the rule. In other words, even as at least two effective vaccines are starting to be distributed, the economic impact of the crisis seems poised to linger for some time to come.

The reality is that all this creates a fertile ground for a dramatic transformation of job markets due to automation and the impact of technology more broadly. History shows that the vast majority of job losses from the adoption of labor-saving technology tend to be concentrated in economic downturns. Routine jobs are especially hard hit, and this largely explains the evaporation of solid middle class jobs and their eventual replacement by less desirable, lower-wage opportunities in the service sector.

Indeed, the economists Nir Jaimovich and Henry E. Siu studied this phenomenon and in a 2018 paper found that "essentially all employment loss in routine occupations occurs in economic downturns."[19] What appears to be happening is that businesses eliminate workers under economic duress; then as the downturn progresses, they incorporate new technology and reorganize workplaces; eventually once recovery occurs, they find that they're able to avoid re-hiring all or most of the workers they previously believed to be essential to their operations. The depth of the current downturn suggests that most businesses will be under enormous pressure to become more efficient, and the longer the crisis lasts, the more time they will have to assimilate new technology—including the latest applications of artificial intelligence—into their business models.

Beyond the purely economic impetus to adopt new technology, the current crisis is unique in that it adds yet another incentive to transition to a more automated workplace. As we saw in Chapter 3, the need for social distancing has already driven a dramatic boost to the adoption of robotic technologies in a variety of areas. Meat packing plants in the United States and elsewhere, for example, have again and again become major nexuses of contagion in settings where hundreds or thousands of workers labor virtually shoulder to shoulder. In environments like this, it's likely inevitable that more automation will be adopted as a way of reducing worker density.[20] While this represents an extreme case, the same is true in virtually every other type of work environment, from factories and warehouses to retail stores to offices. Replacing workers with robots or with smart algorithms translates directly to fewer people in close proximity. Customer-facing service businesses are likely to perceive a marketing advantage in minimizing the direct human interaction that just a few months ago was viewed as a positive rather than a negative. Indeed, this trend is already in play: in

July 2020, the fast food chain White Castle announced that it would begin deploying hamburger cooking robots in order to create "an avenue for reduced human contact with food during the cooking process—reducing potential for transmission of food pathogens."[21] The long-term impact of these factors will probably depend to some extent on the duration of the crisis. As of this writing, however, it seems likely that the situation will persist long enough that at least some of the behavioral changes and customer preferences that have emerged as a result of the pandemic will become ingrained and possibly permanent.

The impact of artificial intelligence on workplaces is not going to be a straightforward narrative of robots stealing jobs. Research has shown that in most cases there is not a one-to-one correspondence between the new technology deployed and an existing job. Rather, it tends to be specific tasks—not entire occupations—that are most susceptible to automation. An influential 2017 analysis by the McKinsey Global Institute found that roughly half of all tasks currently performed by workers globally could, in theory, already be automated using existing technology. McKinsey's analysis showed that only five percent of jobs were at immediate risk of full automation, but that "in about 60 percent of occupations, at least one-third of the constituent activities could be automated, implying substantial workplace transformations and changes for all workers."[22] It's easy to see that if a significant fraction of the tasks being performed by two or three workers can be automated, then there is clear potential for redefining the boundaries between jobs and consolidating the remaining work. It seems very likely that economic pressure along with the need for reduced density in workplaces will create a powerful incentive for many organizations to rethink and reorganize their work environments in order to take advantage of these unrealized efficiencies, and this trend will be amplified by the arrival of far more capable applications incorporating the latest advances in deep learning. The result, in most cases, will be fewer jobs—and those jobs may well be held by different workers with entirely different skillsets and talents.

Aside from the direct automation of jobs and tasks, a second important force is the de-skilling of jobs. In other words, the adoption of new technology allows a role that once required significant skill and experience to instead be filled by a lower-wage worker with little training, or by an interchangeable independent contractor working in the gig economy. A classic example of this is the experience of the famous "black cab" taxi drivers in London. Obtaining a license to drive such a taxi traditionally requires full memorization of virtually all the streets in the city, a laborious process known as acquiring "The Knowledge." The memorization required is so extensive that an analysis by the University College London neuroscientist Eleanor Maguire found that the hippocampus—the area of the brain associated with long-term memory—is, on average, larger in black cab drivers as compared to workers in other occupations.[23] This requirement for prospective drivers to acquire The Knowledge has historically provided a forbidding entry barrier into the profession and thereby ensured cab drivers a solid middle class wage. This has changed dramatically with the advent of GPS and smartphone navigation apps. Now drivers with no knowledge of London streets whatsoever, but access to a smartphone, are able to compete directly, and the onslaught from ride-sharing services and other taxi-like options has had a dramatic and negative impact on the livelihoods of London taxi drivers. In general, de-skilling acts to push down wages by making the job accessible to people will little or no training or experience, while at the same time making workers more interchangeable. This in turn allows businesses to tolerate high turnover rates and further undermines the bargaining power of workers. As both automation and de-skilling progress, there is every reason to expect that inequality will grow and that the fruits of innovation will continue to accrue increasingly to the top of the income distribution.

These technological trends will intertwine with other important ramifications of the pandemic. For example, the wholesale adoption of remote work among white collar workers has decimated the business ecospheres that surround concentrations of office buildings. It seems very likely that the shift toward telecommuting will, at least to some extent, be permanent. Facebook, for example, has already announced that many of its employees will be able to opt for remote work indefinitely.[24] In these once teaming business districts, jobs at restaurants, bars and other businesses that cater to office workers may never return to previous levels. Jobs for the service workers who clean and maintain offices and provide security could likewise be impacted. A second key factor is the likely bankruptcy of a large fraction of the small businesses that disproportionately provide these jobs. By some accounts, up to half of the small businesses that were forced to shut down amidst the pandemic may never reopen.[25] Eventually, the market share once commanded by these small businesses will be captured by larger, more resilient retail and restaurant chains. However, because these larger businesses have greater financial resources and internal expertise, they will be far better positioned to be early adopters of new labor-saving technology. In other words, increasing domination of markets by large enterprises could act directly to accelerate both job automation and de-skilling in the service sector. There's a very real risk that the convergence of all these forces will have a significant dampening effect on the re-generation of the low-wage service jobs that have been a primary engine of American job creation in recent years, and this has the potential to make sustained recovery from the current crisis all the more difficult.

THE COMING WAVE OF WHITE COLLAR AUTOMATION . . .
AND WHY TEACHING EVERYONE TO CODE IS NOT A SOLUTION

The specter of job automation typically conjures up images of industrial robots toiling in factories or warehouses. The conventional wisdom suggests that, while lower-wage, less-educated, blue collar workers face a dire threat from technology, knowledge workers educated with at least a bachelor's degree—or in other words, anyone whose occupation consists of tasks that are primarily of an intellectual, rather than a manual, nature— remain on relatively safe ground. The reality, however, is that white collar jobs, and in particular those that center on relatively routine analysis, manipulation, extraction or communication of information, are going to be squarely in the sights as artificial intelligence advances and is deployed ever more widely.

Indeed, in many cases, white collar professionals in information-oriented jobs may prove to be even more vulnerable to displacement by technology than less educated workers in occupations that require physical manipulation of the environment. This is because automation of these roles requires no expensive machinery and there are no challenges to overcome in areas like machine vision or robotic dexterity. Instead, eliminating many of the tasks that occupy the time of these workers requires only sufficiently powerful software. And the incentive to eliminate white collar work is further amplified by the fact that these higher-skilled workers are generally paid significantly more than their blue collar counterparts. As we've already seen, nearly half of recent college graduates are underemployed, and this is likely driven to some degree by the impact of technology on the more routine entry-level positions that traditionally have offered the first rung on the ladder to professional success.

Though the jobs most at risk will continue to be those that are more routine, it's important to realize that the line between tasks that can be automated and those that are perceived as safe is certain to be dynamic and will shift relentlessly to encompass ever more activities as artificial intelligence continues to advance. Previously, the automation of a knowledge-based activity would have required a computer programmer to lay out a step-by-step procedure, with each action and decision explicitly articulated. This tended to limit software automation to truly routine and repetitive undertakings, often in clerical areas like general book-keeping or accounts payable and receivable. The rise of machine learning, however, means that algorithms are now turned loose to essentially write their own computer programs by churning through reams of data and finding patterns and interrelationships that are often beyond direct human perception. In other words, it is the essence of machine learning to transform tasks that were once perceived as inherently non-routine into activities that are now susceptible to automation.

There are already numerous examples of how software automation, often incorporating machine learning, is beginning to encroach on activities across a wide variety of white collar occupations. In the field of law, for example, smart algorithms now review documents to determine if they need to be included in the legal discovery process, and artificial intelligence systems are becoming increasingly adept at legal research. Predictive algorithms analyze historical data and assess the probability of everything from the outcome of cases before the Supreme Court to the likelihood that a particular contract might some-day be breached. In other words, AI is already beginning to im-pact judgment-driven activities that would have once been the purview of only the most experienced attorneys. Large media organizations increasingly rely on systems that automate basic journalism by analyzing a stream of data, identifying the story it contains and then automatically generating narrative text. Com-panies like Bloomberg use these systems to create news articles covering corporate earnings reports almost instantaneously. As the ability of artificial intelligence to handle natural language improves, it's likely that nearly any type of routine writing intended for either internal or external organizational communication will be increasingly susceptible to automation. Analytical jobs in industries like banking and insurance are likely to be especially vulnerable to all of this. A 2019 report from Wells Fargo, for example, predicted that about 200,000 jobs in the U.S. banking industry would evaporate as the result of advancing technology over the next decade.[26] The impact of automation on Wall Street is already evident, and what were once bustling and chaotic trading floors are now largely filled with the soft hum of machines. By 2019, the major stock exchanges were staffed by only small groups of people relegated to certain areas of the trading floor.[27] The coronavirus pandemic has demonstrated that even these few holdouts are no longer essential, as the exchanges moved rapidly to fully electronic trading.

Call centers that provide customer service or technical support are another area that is clearly ripe for disruption. Rapid advances in the natural language processing capabilities of artificial intelligence are producing applications that can automate ever more of this work via both voice communication technology and online chatbots. These jobs, of course, were already highly vulnerable to offshoring. As technology has advanced, however, many of the call center jobs in lower-wage countries like India and the Philippines are being vaporized by automation. Responding to customer service queries is a task that in many ways is ideally suited to machine learning. Each interaction between a customer and a call center worker generates a rich set of data, including the question asked, the answer provided and whether the interaction fully resolved the issue. Machine learning algorithms can churn through thousands of these interactions and quickly become proficient at responding to the significant fraction of queries that tend to come up again and again. And once the system is in place, and as more customer calls come in, the algorithms get smarter and smarter. There are literally dozens of startup companies offering AI-powered chatbots to automate customer service. Many of these are positioned in specific sectors such as healthcare or financial services.[28] As these technologies continue to advance, call center staffing is likely to fall as things eventually reach a point where a human operator is required only for the most challenging customer interactions.

The ability to write computer code is often presented as a kind of panacea for technological job market disruption. Those losing jobs in industries like journalism or even coal mining have been advised to "learn to code." Coding academies have sprung up, and there have been many proposals to make computer programming classes mandatory in high school, or even earlier. In truth, however, writing computer code is certain to be subject to the same forces that will disrupt other types of white collar work. As with call centers, outsourcing is often the leading edge of automation, and much routine software development is already offshored to lower-wage countries, especially India. Nearly all the major tech companies have been making significant investments in tools that automate computer programming. Facebook, for example, has developed a tool called Aroma that works as a kind of AI-powered "autocomplete" for computer programming that leverages a huge database of public domain computer code.[29] DARPA has also funded research into automating the development, debugging and testing of computer code. Even OpenAI's GPT-3, a general language generation system trained on a vast number of documents extracted from the internet, is able to complete some routine programming tasks.[30]

The bottom line is that though learning to program a computer may certainly be a useful and rewarding undertaking, the

days when acquiring that skill guarantees a decent job are coming to a close. And the same will be true for a wide spectrum
of other white collar occupations. As technology begins to encroach on even these more educated and highly paid workers,
inequality is likely to become ever more top heavy, with a tiny
elite that owns vast amounts of capital pulling away from everyone else. As better-paid workers are increasingly impacted, this
will further undermine consumer spending and the potential for
robust economic growth. One possible upside, however, is that
better-paid knowledge workers wield far more political power
than their counterparts who work in factories or low-wage service occupations. As a result, the impact on white collar jobs
may actually help to galvanize support for a policy response to
technological disruption of the job market.

WHICH JOBS ARE SAFEST?

Over the past few years, I've traveled to nearly every continent
and given dozens of presentations on the potential impact of
artificial intelligence and robotics on job markets. Regardless of
the country I happen to be in, I've found that the most common
questions I receive from the audience are nearly always the same:
what jobs are most likely to be safe, and what fields should I
advise my children to study? The general answer is perhaps a
bit obvious and unsatisfying: avoid jobs that tend to be fundamentally routine and predictable in nature. These are clearly
the areas that will see the most significant near-term impact of
AI-powered automation. Another way to phrase this might be
"avoid jobs that are boring." If you're coming to work and facing new challenges every day and if you're constantly learning
on the job, then you're probably well positioned to stay ahead
of technology, at least for the foreseeable future. If, on the other
hand, you spend a lot of time cranking out the same kinds of

reports, presentations or analyses again and again, you should probably start to worry—and begin to think about adjusting your career trajectory.

More specifically, I think the jobs least susceptible to automation in the near to intermediate term fall into three general areas. First, jobs that are genuinely creative in nature are likely to be relatively safe. If you're thinking outside the box, coming up with innovative strategies to solve unforeseen problems or building something genuinely new, then I think you will be well positioned to leverage artificial intelligence as a tool. In other words, the technology is much more likely to compliment you than it is to replace you. To be sure, significant research into building creative machines is underway, and AI will inevitably begin to encroach on creative work as well. Already smart algorithms can paint original works of art, formulate scientific hypotheses, compose classical music and generate innovative electronic designs. DeepMind's AlphaGo and AlphaZero have injected new energy and creativity into professional Go and chess competitions because the systems represent truly alien intelligences, often adopting unconventional strategies that astonish human experts. However, I think that for the foreseeable future artificial intelligence will be used to amplify, rather than replace, human creativity.

A second safe area consists of those jobs that put a premium on building meaningful and complex relationships with other people. This would include, for example, the kind of empathetic, caring relationship that a nurse might have with a patient, or that a business person or consultant offering sophisticated advice might develop with a client. It's important to note that I'm not referring so much to short-term service encounters that involve smiling and being friendly with customers, but rather those that require deeper and more complex interpersonal interactions. Once again, AI is also encroaching into this area; as we saw in Chapter 3, chatbots can, for example, already provide rudimentary mental health therapy, and there will continue to be significant advances in AI's ability to perceive, respond to and simulate human emotion. I think it will be a long time, however, before machines become capable of developing truly sophisticated, multidimensional relationships with people.

The third general category of safe jobs includes occupations that require significant mobility, dexterity and problem-solving skills in unpredictable environments. Nurses and elder care assistants would also fall into this category, as would skilled trade occupations like plumbers, electricians and mechanics. Building affordable robots capable of automating work of this type likely lies far in the future. These skilled trade jobs will generally represent some of the best opportunities for those who choose not to pursue a college education. In the United States, I think we should be putting far more emphasis on accessible vocational training or apprenticeships that prepare young people for these opportunities rather than simply pushing ever more high school graduates to attend college or university.

The most important factor, however, may not be so much which occupation you choose but how you position yourself within it. As artificial intelligence advances, it's likely that across broad swathes of the job market, jobs consisting largely of routine "nuts and bolts" activities will evaporate, while those who focus in areas requiring creative skills or who can leverage extensive professional networks in ways that add value to organizations will rise to the top. In other words, there's likely to be something of a winner-take-all or superstar effect of the kind you see among athletes or entertainers imposed on occupations that were previously more homogeneous in terms of opportunity. A lawyer with strong courtroom skills or the client connections that bring business to the firm will likely continue to do well even as artificial intelligence advances. An attorney who, on the other hand, toils away mostly on legal research or contract analysis may be in a less promising situation.

The best way for you as an individual to adapt to this situation is probably to select an occupation that you genuinely enjoy—something that you're passionate about—because this will increase your chances of excelling and becoming an outlier in the field. Going forward, choosing an occupation simply because the field has traditionally provided lots of jobs may not be such a good bet. The problem, of course, is that this may be good advice for a given individual, but it is not a systemic solution. Many people will, in all likelihood, be left behind as these transitions unfold, and ultimately, I think we will need policies to address that reality.

THE ECONOMIC UPSIDE

Though the potential impact of artificial intelligence on the job market and on economic inequality are real concerns, there is also no doubt that the technology is poised to deliver enormous benefits across both the economy and society. Increased automation will boost the efficiency of production and lead directly to lower prices for goods and services. In other words, AI will be a critical tool for alleviating—and eventually eliminating—poverty by making all the things that people need to thrive more abundant and affordable. Artificial intelligence deployed in research, design and development will lead to entirely new products and services that otherwise might have been unimaginable. New medicines and treatments will lead to dramatic economic benefits while increasing the well-being of nearly everyone.

Two reports released in late 2018, one from the McKinsey Global Institute[31] and the other from the consulting firm PwC,[32] both make a strong argument that artificial intelligence will deliver a massive boost to the global economy by the year 2030.

McKinsey's analysis projects that AI will add about $13 trillion to economic output worldwide, while PwC's estimate comes in at $15.7 trillion. In other words, AI is likely poised to add new global economic value roughly equivalent to China's current $14 trillion GDP over the next decade or so. McKinsey's analysis suggests that these gains will arrive in a fashion that traces an S-curve—"a slow start due to substantial costs and investment associated with learning and deploying [artificial intelligence], but then an acceleration driven by the cumulative effect of competition and an improvement in complementary capabilities."[33] By 2030, it's likely we'll find ourselves on the steep, accelerating portion of the curve with both the technology and the economic gains associated with it progressing rapidly.

These estimates largely fail to capture the most dramatic benefits from artificial intelligence over the long term. As I argued in Chapter 3, the single most important promise of AI is that it can help us escape from our age of technological stagnation. If artificial intelligence allows us to jump-start innovation across a broad range of scientific, engineering and medical fields, the potential return on our investment will be staggering. What is perhaps most important is the critical need to amplify our collective intelligence and creativity in ways that will allow us to address the daunting challenges that are sure to confront us—including everything from climate change to new sources of clean energy to managing the next pandemic. These things are difficult to quantify with economic analysis, but I would argue they alone make artificial intelligence an indispensable tool that we simply cannot afford to leave on the table, even as it comes coupled with unprecedented economic and social risks.

The key challenge before us is to find ways to address downsides like technological unemployment and increased inequality while continuing to invest in AI and fully leverage the advantages the technology will bring. The fundamental economic challenge we will face is one of distribution. The potential economic gains associated with artificial intelligence are undeniable, but there is absolutely no guarantee that these benefits will be shared broadly or fairly across the population. Indeed, if we take no action at all, it seems a near certainty that the gains will accrue overwhelmingly to a small sliver of people at the top of the income distribution, while the bulk of the population will be largely left behind or potentially even made worse off. And as we've seen, this in turn could erode broad-based consumer demand and dampen down both productivity gains and economic growth. In other words, a failure to address the economic downsides of AI may well limit the full realization of the technology's upside. Avoiding this default outcome will, I think, require dramatic and unconventional policy initiatives. Traditional solutions that have been deployed for decades—job retraining programs or pushing ever more people to attend college—are unlikely to be sufficient, especially given that artificial intelligence is already having a significant impact on higher-skill jobs, and this trend will only gain traction as the technology becomes ever more capable.

FIXING THE DISTRIBUTIONAL PROBLEM

In my view, the most straightforward and effective way to address the distributional challenge brought about by artificial intelligence advances is simply to give people money. In other words, supplement the incomes of all or the bulk of the population with some version of a guaranteed minimum income, negative income tax or basic income. The idea that has recently gained the most traction is an unconditional universal basic income, or UBI. The visibility of UBI as a policy response to AI-driven automation was dramatically accelerated in 2019 by the presidential candidacy of Andrew Yang. Yang, who sought the

Democratic nomination for president, ran primarily on a platform of a $1,000 per month "Freedom Dividend" that would
be paid to all U.S. citizens. His campaign gained remarkable
traction largely as the result of a vibrant online following, and
his participation in the Democratic debates pushed UBI into the
mainstream and exposed large numbers of Americans to the
idea for the first time.

One of the primary advantages of an unconditional basic
income is that, because it is paid to everyone regardless of employment status, it doesn't destroy the incentive for recipients
to work or engage in entrepreneurial activity that generates additional income. In other words, it avoids one of the biggest
problems with traditional safety net programs: the tendency to
create a poverty trap. Because programs like unemployment insurance or welfare payments are phased out or eliminated entirely once the recipient finds a job and begins to earn income,
there can be a powerful disincentive to seek employment. Ac-
cepting even a low-paying job puts the person's existing income
at immediate risk. As a result, people often get trapped into
dependency on safety net programs and see little in the way of
a concrete incentive to take small steps toward a better future.
A universal basic income, in contrast, is unaffected by employment, and therefore anyone who chooses to work or perhaps
start a small business that generates extra income will always
be better off than the person who simply sits at home and collects the monthly UBI payment. The UBI creates an absolute
income floor, but there always remains a strong incentive to
earn more. Despite this advantage, there is, for many people,
a strong psychological aversion to the idea of simply handing
money to people, or as some have put it, "paying people to be
alive," and this attitude is likely to continue to create a significant political impediment to actual implementation of a UBI.

To be sure, there are other policy alternatives, one of the most commonly cited being a jobs guarantee. The idea of having the government become the employer of last resort for anyone who needs a job may seem superficially attractive, but I think it has significant disadvantages. A jobs guarantee would be far less universal than a basic income, and inevitably many of the people who are most in need of assistance would be left out. Such a system would require a massive, costly and likely ever-expanding bureaucracy. Managers would need to ensure that workers actually show up to do whatever job they've been assigned, and there would doubtless be a whole slew of disciplinary issues ranging from absenteeism to poor performance to "me too" situations. Any policies geared toward disciplining or terminating workers who failed to meet the specified standards would be fraught with controversy and quite possibly with accusations of discrimination or unequal treatment. Ultimately, the government would either have to fire those who underperformed or broke the rules—which would exclude any impacted individuals from the safety net—or the jobs program would effectively become the equivalent of a very expensive and inefficient basic income scheme. A large fraction of the positions created would in all likelihood be "bullshit jobs," and unlike a basic income program, a jobs guarantee would directly attract workers away from more productive positions in the private sector. In contrast, a basic income requires little in the way of a bureaucracy and would take advantage of the government's existing competence at sending out checks via programs like Social Security.

While I think a basic income is ultimately the best overall solution to the distributional issues that will emerge as artificial intelligence becomes ubiquitous, it is by no means a panacea. Rather, I view UBI as a foundation upon which to build a more effective and politically palatable solution. The most important problem is that though a basic income delivers money into the hands of people, it alone does not replicate the other important qualities associated with a traditional job. A meaningful job provides a sense of purpose and dignity. It occupies time and creates an incentive to work hard and to excel in the hope of obtaining a raise or a promotion. The desire to obtain a good job is also a critically important incentive for individuals to pursue further education and training.

I believe it's possible to modify a basic income program so it, at least in part, replicates some of these qualities. Since the publication of my first book, *The Lights in the Tunnel: Automation, Accelerating Technology and the Economy of the Future,* in 2009, I've advocated for a basic income scheme that directly incorporates incentives. Though everyone should receive some minimal guaranteed payment, I think there should also be opportunities to earn somewhat more by pursuing certain activities. The most important incentive by far should be to pursue further education. Imagine a world where everyone receives exactly the same monthly UBI payment beginning at age eighteen or perhaps twenty-one. In this situation, a high school student at risk of dropping out of school might see very little reason to work hard to obtain that diploma. The monthly check, after all, is going to be the same no matter what. And if—as it seems to already be the case—getting that diploma will not be sufficient to obtain a good job anyway, then why stay in school? I think this kind of disincentive would be disastrous and would raise the specter of a less-educated population even as we face a future that is becoming vastly more complex and fraught with difficult challenges and trade-offs. Therefore, why not simply pay a bit more to those who graduate from high school? This idea of building incentives into a basic income program might be expanded to include more advanced education and perhaps other areas such as community service work. The ultimate vision is to create opportunities that would offer people meaningful ways to spend their time and achieve a sense of accomplishment. Perhaps most importantly, those who act on the incentive to pursue further education will increase the odds that they will be able to access still more opportunities through employment or entrepreneurial activity. It's likely that as artificial intelligence is deployed ever more widely, it will provide powerful tools that individuals will be able to leverage to start a small business or perhaps generate income through freelancing opportunities, but taking advantage of these opportunities will require achieving at least a minimum educational threshold. Preserving a strong incentive for everyone at every level of our society to strive for the highest level of education within their ability should be one of our most important goals.

Another major problem with UBI is that it's very expensive. Distributing income unconditionally to every adult American will cost trillions, and voters are likely to recoil at the idea of sending monthly checks to the already well-off. I think there may be opportunities to effectively phase out the UBI at higher incomes, without impacting the incentive to work. The best way to do this might be to means-test a UBI against only "passive income." If you already have significant income that comes to you automatically, without any need for work or action on your part—if you receive a pension, Social Security or substantial investment income—then I think it would be reasonable to phase out or eliminate a UBI payment accordingly. Active income that results from work or direct management of a business would not affect the UBI, except perhaps at a very high income level. Many would perceive this as unfair, but the idea behind a basic income is, after all, to provide everyone with at least a minimal guaranteed income floor. If you already have access to such a payment, then arguably you don't need the UBI. No policy initiative is ever going to make the world completely fair. The best we can realistically hope for is a program that moderates inequality, eliminates the most dire forms of material deprivation and ensures that consumers have the income they need to continue driving economic growth.

All these ideas, of course, face challenges of their own. If we incorporate incentives into a basic income scheme, then who exactly gets to decide what those incentives are? For many people, this will immediately raise the specter of an overbearing nanny state undermining freedom of choice and injecting itself into the daily fabric of our lives. Still, I think it ought to be possible to come to some broad agreement on at least a minimal set of incentives that are clearly advantageous both for individuals and for society as a whole, and once again, I think pursuit of education stands apart as clearly being the most important. A related concern would be the politicization of a basic income program. It's quite easy to imagine a future where nearly every politician runs on a platform of "I will increase your monthly UBI payment." For this reason I think it would make a lot of sense to remove the administration of a basic income program from the political process and place it in the hands of a dedicated technocratic agency operating according to clear guidelines—in other words, an institution similar to the Federal Reserve.

None of this is to suggest that we should abandon more conventional solutions to unemployment, underemployment or rising inequality. We should do everything possible to ensure that the maximum number of workers are able to successfully transition as the impact from artificial intelligence and robotics accelerates in the coming years and decades. In particular, we should invest in community colleges and affordable vocational training or apprenticeship programs that offer an alternative to the predatory for-profit schools that, in the United States, currently occupy much of this space. Still, I think that eventually the disruption will be of such magnitude that programs of this type will ultimately fall short, and we will need to adopt more unconventional solutions.

The political hurdles standing in the way of a basic income will remain daunting, and I think that realistically such a program may need to begin at a minimal level and then gradually be ramped up over time. Before a national program can be enacted, we need more data and more actual experience with UBI; therefore, we should initiate experiments designed to find the optimal policy parameters. I hope that some of these experiments might eventually include my idea of incorporating incentives. The data generated through basic income experiments will allow us to craft a program that will scale effectively and help ensure broad-based prosperity in a future shaped increasingly by AI.

THE POTENTIAL FOR technological unemployment and increasing inequality is just one of the major concerns that comes coupled with the rise of artificial intelligence. The next two chapters will focus on a range of other dangers that are already becoming apparent or are likely to arise as the technology progresses.

CHAPTER 7

CHINA AND THE RISE OF THE AI SURVEILLANCE STATE

THE XINJIANG AUTONOMOUS REGION CONSTITUTES CHINA'S northwestern frontier. The region is massive—roughly two and half times the size of Texas—and borders seven nations aside from China: Mongolia to the northeast, Russia to the north, and Kazakhstan, Kyrgyzstan, Tajikistan, Afghanistan, Pakistan, and India to the west. The climate and the terrain are formidable—primarily rugged mountains and desert punctuated by the oasis cities where most of the province's twenty-four million people are clustered. The legendary Silk Road—or really "roads," as it was actually a network of routes—cut across Xinjiang and made the region central to the East-West trade that contributed to the rise of civilizations throughout Eurasia. Marco Polo traveled this route in the late thirteenth century and would have encountered bustling bazaars and laden camels not unlike those that can be seen in Xinjiang today.

Xinjiang has been thrust into the limelight not for its rich history, but rather because an Orwellian future has been imposed on the region's largest ethnic group, the Uyghurs. In cities like Kashgar, China has built an oppressive surveillance state,

201

powered by a combination of massive police presence, physical checkpoints and advanced technology. Virtually everyone in the city is continuously watched: thousands of cameras line the streets, mounted on buildings and clustered on telephone poles. As residents move through the city, they are stopped at checkpoints and allowed to advance only after being identified by facial recognition systems.[1]

While Xinjiang is ground zero for China's surveillance initiatives, the region also acts as a proving ground for techniques and technologies that are gradually being deployed across the entire country. China is expected to have installed nearly 300 million cameras by 2020, many of which are linked to facial recognition technology or feature other AI-driven tracking techniques, such as identifying pedestrians based on their gait or clothing.

In Xinjiang, Uyghurs who deviate from prescribed behaviors or engage with forbidden ideas, such as reading the Koran, risk being sent to one of the massive "re-education camps" that China has built in the region. Even in other areas of the country, the Chinese government has a terrifying vision for systemic behavior modification, implemented through the deployment of a comprehensive social rating system. Eventually, nearly all aspects of a person's life—consumer purchases, physical movements, social media interactions and associations with others—will be surveilled, recorded and analyzed. This information will then be used to generate an overall social rating for each individual. Those who score low on this metric will suffer penalties such as being barred from public transportation or having their children prevented from enrolling in schools.

All this is being accelerated by China's rapid rise to become a world leader in artificial intelligence research and development. By some measures, such as the sheer number of computer scientists and engineers working in the field and the volume of research papers published, China has already taken a lead over the United States. The country is investing massively and has made artificial intelligence a strategic national imperative. Its leaders appear to be both engaged and knowledgeable. In early 2018, Chinese president Xi Jinping gave a televised address from his office, and books on AI and machine learning were spotted in the background.[2] The government is also helping to fund hundreds of startup companies, many of which are valued at billions of dollars and are clear technology leaders.

As China assumes its role as one of the world's two primary centers of artificial intelligence research and development, the ongoing competition in this arena with the United States and the West is likely to become ever more intense. A large fraction of China's emerging AI industry is focused on developing facial recognition and other surveillance technologies, and these companies are finding eager customers, not just in China but also in countries throughout the world. And, as we'll see, AI-based surveillance technologies are by no means limited to authoritarian regimes. Facial recognition, in particular, is being widely deployed in the United States and other democratic countries and has already led to intense debate and accusations of bias and misuse. These issues will become only more fraught as the technology continues to become more powerful and—unless it is strictly regulated—ubiquitous.

CHINA'S LEAP TO THE FOREFRONT OF ARTIFICIAL INTELLIGENCE RESEARCH AND DEVELOPMENT

In June 2018, a major conference on computer vision was held in Salt Lake City, Utah. In the six years since the famous 2012 ImageNet competition, the field of machine vision had advanced dramatically, and researchers were now focused on solving far more difficult problems. One of the highlights of the conference was the Robust Vision Challenge. This competition, sponsored by major companies including Apple and Google, pitted teams from universities and research labs across the world against each other in a series of challenges geared toward reliably identifying images in varying situations, such as indoor or outdoor lighting or differing weather conditions.[3] This capability is crucial to applications like self-driving cars or robots that operate in varied environments. One of the most important segments of the contest focused on stereo machine vision: using two cameras in much the way that we employ our eyes. By interpreting visual information from slightly different vantage points, our brains are able to generate a three-dimensional representation of a scene. Two properly positioned cameras allow an algorithm to do something similar.[4]

The winning team took many people by surprise: a group of researchers from China's National Defense Technology University. The university was founded in 1953 as the People's Liberation Army (PLA) military academy of engineering and has received numerous national awards for research and innovation, especially in computer science. According to its website, the university "bases its educational efforts on the Party's innovation theory to cultivate loyal and qualified successors."[5] That seems like a pretty good indication that there is, at best, a very porous dividing line between academic or commercial AI research in China and the country's political, military and security apparatus.

It is, of course, routine for the Chinese government to intervene in and exert some degree of control over nearly every aspect of the country's economy and society. However, China's recent rapid progress in artificial intelligence has been significantly accelerated and orchestrated by an explicit industrial policy articulated by the central government.

Many observers believe that the catalyst for the sudden surge of interest in AI on the part of the Chinese Communist

Party was the highly touted contest between DeepMind's AlphaGo system and Go champion Lee Sedol that took place in March 2016. The game of Go originated in China at least 2,500 years ago and is wildly popular and revered among the Chinese public. AlphaGo's 4–1 triumph, which took place over seven days in Seoul, South Korea, was viewed live by more than 280 million people in China—nearly three times the audience that tunes in for a few hours to watch a typical Super Bowl. The specter of a computer defeating a top human player at an intellectual pursuit so deeply rooted in Chinese history and culture made an indelible impression on the public as well as on Chinese academics, technologists and government bureaucrats. Kai-Fu Lee, a Beijing-based venture capitalist and author, calls the AlphaGo–Lee Sedol match "China's Sputnik moment," in reference to the Soviet satellite that galvanized public support for the U.S. space program in the 1950s.[6]

Just over a year later, a second contest was held in Wuzhen, China. In a three-game match carrying a $1.5 million prize for the winner, AlphaGo defeated the Chinese player Ke Jie, who was then ranked number one in the world, by prevailing in three straight games. This time around, however, there was no live audience. The Chinese government, perhaps anticipating the outcome, had issued a censorship order forbidding any live broadcast, or even live text-based commentary, of the match.[7]

Two months after Ke Jie's loss to AlphaGo, in July 2017, the Chinese central government released an explicit plan designating artificial intelligence as a strategic national priority. Entitled "New Generation Artificial Intelligence Development Plan," the document declared that AI was poised to "profoundly change human society and life and change the world" and then went on to lay out an extraordinarily ambitious step-by-step course toward domination of the technology by the year 2030. By 2020, the plan's authors wrote, China's "overall technology and application of AI will be in step with globally advanced levels" and "the AI industry will have become a new important economic growth point." Next, "by 2025, China will achieve major breakthroughs in basic theories for AI, such that some technologies and applications achieve a world-leading level and AI becomes the main driving force for China's industrial upgrading and economic transformation." And finally, "by 2030, China's AI theories, technologies, and applications should achieve world-leading levels, making China the world's primary AI innovation center, achieving visible results in intelligent economy and intelligent society applications, and laying an important foundation for becoming a leading innovation-style nation and an economic power."*,[8]

---

* If you have any doubts about the power of deep neural networks when applied to language translation, compare these two introductory sections from China's "New Generation Artificial Intelligence Development Plan." One is a Google machine translation of the original Chinese government document. The other was professionally translated by a team of four linguists.

The first paragraph from each document is below. Can you tell which is which?

A. The rapid development of artificial intelligence will profoundly change human society and the world. In order to seize the major strategic opportunities in the development of artificial intelligence, build the first-mover advantage in the development of artificial intelligence in China, and accelerate the construction of an innovative country and a world power of science and technology, this plan was formulated in accordance with the deployment requirements of the Party Central Committee and the State Council.

B. The rapid development of artificial intelligence (AI) will profoundly change human society and life and change the world. To seize the major strategic opportunity for the development of AI, to build China's first-mover advantage in the development of AI, to accelerate the construction of an innovative nation and global power in science and technology, in accordance with the requirements of the CCP Central Committee and the State Council, this plan has been formulated.

The answer is that B is the human-translated version. (See endnote 8.)

The publication of this document was crucial not because the Chinese central government has the ability to directly micromanage the development of artificial intelligence capability throughout the country, but because it defined an overall strategy and, perhaps more importantly, created clear incentives for regional and local governments. In China's system, a great deal of power is delegated to the Communist Party officials who run the country's various regions and cities. Advancement through the party's ranks is largely meritocratic, and an official's career trajectory is heavily dependent on how he or she performs in a competitive ecosphere that values performance as measured by specific metrics. For those who manage to stand out, there are potentially few limits. Xi Jinping spent much of his career as a top official in the Fujian and Zhejiang provinces and later in the city of Shanghai.

Even before the central government explicitly embraced artificial intelligence, specific regions were already making substantial investments and encouraging AI startup companies. Much of this was focused in high-tech corridors such as the southern city Shenzhen and the Zhongguancun area in northwestern Beijing, which is close to the country's two most prestigious universities, Peking and Tsinghua, and is often referred to as "China's Silicon Valley." However, publication of the strategy document in 2017 effectively created an explicit AI metric upon which regional officials knew they would likely be judged. As a result, regions and cities across the country quickly jumped into the fray, creating special economic zones and startup incubators and providing direct venture capital and rent subsidies to AI startups. The investments made by a single city can easily reach billions of dollars. This kind of loosely coordinated top-down directive with a focus on innovation would be hard to imagine in the United States. The American version of interregional competition is generally a far more zero-sum phenomenon—Texas luring businesses from California or cities offering large tax breaks to big companies in return for job-creating facilities.

China enjoys a number of critical advantages as it powers forward in artificial intelligence. Many of these advantages derive directly from the country's massive population. As of March 2020, China had about 900 million active internet users, more than the United States and Europe combined and roughly one fifth of the total number of people online globally.[9] Internet access, however, had been extended to only about sixty-five percent of its population, versus ninety percent in the United States.[10] In other words, China has far more online growth potential. Among China's 1.4 billion people, there are an enormous number of smart and ambitious high school and university students who are eager to become proficient in technologies like deep learning in the hope of eventually joining—or launching—one of the exploding number of Chinese AI startups, many of which have achieved valuations over a billion dollars. These young people are among the most dedicated and enthusiastic participants in online courses offered by top U.S. universities like MIT and Stanford. They also eagerly comb through technical papers published by top AI researchers in North America and Europe. As a result, China is quickly developing a large pool of talented and extremely hard-working engineers who are continuously hoovering up state-of-the-art knowledge generated in the West and will soon be poised to leverage AI across virtually every dimension of China's economy and society.

The most important advantage, however, lies in the sheer volume, as well as the type, of data that Chinese economic activity generates. As a developing country, China has far less invested in legacy infrastructure, and as a result, the country has leapt directly to the very frontier of mobile technology. The Chinese public uses smartphones in a vastly wider array of activities than is typical in the West. All this is driven especially by the popularity of Tencent's WeChat app. Introduced in 2011, WeChat has gained overwhelming popularity in China and also among the Chinese diaspora in other countries.

At its core, WeChat is a messaging app roughly comparable to Facebook's WhatsApp. However, Tencent made a decision early on to dramatically expand the capability of WeChat by allowing third parties to add their own functionality using what are called "official accounts." These essentially amount to mini apps and are extraordinary popular with businesses of all types, especially when combined with WeChat's ability to make digital payments. In the United States and in other western countries, the norm is for every business to have its own mobile app. In China, WeChat has evolved into a kind of "master app as platform," and millions of businesses and organizations use it to interface with the public. The Chinese use WeChat not just to communicate, but also to pay bills at restaurants, to book doctor appointments, for online dating, to pay their utility bills, to hail taxis and, in essence, to do just about everything else. And the number of services available through WeChat expands continuously. Unlike systems such as Apple Pay, which require merchants to invest in expensive point-of-sale equipment, mobile payment via WeChat can be implemented simply by displaying a barcode for customers to scan. As a result, even the smallest businesses, such as street food vendors, can easily accept digital payments. Throughout China, WeChat payment is far more popular than credit cards and is even displacing cash in many venues.

The upshot is that there is vastly more digital activity in China, and it extends to a far greater depth in the overall economy—capturing a torrent of transactions that, in the United States or Europe, would likely be offline. And every payment, every booking, every taxi ride and every interaction of any type generates data that is ideally suited to be gobbled up by deep learning algorithms.

In addition to being more abundant, data is also generally far more accessible to AI entrepreneurs in China. Though data privacy regulations do exist, they are nowhere near as stringent as in the United States or especially in Europe. Nor does the public tend to be particularly focused on most of these issues. Concerns about personal privacy, or perhaps possible racial bias in algorithms—issues that can quickly generate incandescent outrage in democratic societies—are nonexistent or barely cause a ripple in China. While Google's access to NHS data that was originally contracted to DeepMind immediately led to an outcry in the United Kingdom, Chinese tech companies generally benefit from a smoother path to implementation and profitability when it comes to leveraging artificial intelligence in areas like healthcare and education. If data is the new oil, then China's AI entrepreneurs are new age wildcatters—drilling and erecting pumps to extract value at every promising location across a relatively unpoliced digital terrain.

Even before the explosion of venture-backed AI startups, China's major technology companies, especially Tencent, Alibaba and Baidu, were making massive investments in artificial intelligence research and development. Baidu, which is often called "the Google of China" and is the country's leading internet search engine, has developed deep expertise in areas like speech recognition and language translation, but is also pushing aggressively into other areas. In 2017, for example, Baidu introduced Apollo, an open-source autonomous vehicle platform—essentially a kind of "Android for self-driving cars"—which the company gives away freely to companies within China's highly fragmented automotive manufacturing industry.[11] Global car companies, including BMW, Ford and Volkswagen, as well as technology providers like NVIDIA, have also signed on as partners. In return, Baidu gets access to the data that is generated by the vehicles, which it can then use to train its algorithms.

In other words, Baidu is following a unique strategy that may eventually offer advantages similar to what Tesla enjoys with its hundreds of thousands of camera-equipped cars.

Much of the early Chinese AI progress was driven significantly by transfers of knowledge and talent from the United States and other Western countries. American researchers with proficiency in the Chinese language, in particular, have been targeted for recruitment. In 2014, for example, Baidu hired one of the U.S.'s highest-profile deep learning experts, Andrew Ng, who was then leading the Google Brain project, the first initiative to leverage large-scale deep neural networks at Google. Ng, who stayed with the company for three years before returning to Silicon Valley, set up Baidu's primary artificial intelligence research lab in Beijing. Then in 2017, Baidu hired Qi Lu, a top AI executive at Microsoft, to be the company's chief operating officer.[12] Lu, who holds a PhD from Carnegie Mellon University, is one of a growing number of immigrants, educated at top American graduate programs, who are choosing to return to China because the business opportunities centered on AI are often perceived as more attractive. Indeed, the abundant opportunities and rapidly shifting terrain often lead to high turnover among talented Chinese AI experts. Lu remained with Baidu for only about a year and now runs a startup incubator in Beijing.

Access to research and algorithms developed in the West has also played a key role. About a year after AlphaGo defeated Ke Jie, Tencent announced that its own Go-playing software, Fine Art, had also succeeded in defeating the Go master. Tencent's system, however, was likely heavily inspired by, or perhaps even directly copied from, DeepMind's published work. Most of the Western AI researchers with whom I spoke don't seem particularly concerned about this kind of knowledge transfer or view progress in terms of national competition; they believe strongly in a global system that emphasizes open publication of research and a free exchange of ideas. When I asked DeepMind CEO Demis Hassabis about a perceived "AI race with China," he told me that DeepMind publishes openly and that he knows "Tencent has created an AlphaGo clone," but that he doesn't view it as "a race in that sense because we know all the researchers and there's a lot of collaboration."[13]

Also, by all accounts, Chinese researchers have been making substantial contributions to the published body of AI research. According to an early 2019 analysis by the Allen Institute for AI, China had already surpassed the United States in terms of total research papers on artificial intelligence published as far back as 2006.[14] Because there is a general consensus that many of these papers are of relatively low quality or report very incremental progress, the Allen Institute did a further analysis focused on the smaller number of published papers that were heavily cited by other researchers. The analysis found that, assuming the continuation of current trends, China would have passed the United States in terms of publication of papers in the top fifty percent as measured by citations by the end of 2019 and the top ten percent most cited papers by 2020. Chinese researchers were on track to publish more truly elite papers—ranking in the top one percent in terms of citations—than the U.S. by 2025. By still another metric, China is also already ahead of the United States in terms of the total number of artificial intelligence patents filed.

Not everyone buys into the idea that China is on the verge of surpassing the United States in artificial intelligence research and development. Jeffrey Ding, a researcher at the Centre for the Governance of AI at Oxford University's Future of Humanity Institute, conducted an analysis in 2018 that rated AI capability in both the U.S. and China according to four metrics: the installed base of AI computing hardware, availability of data suitable for machine learning, proficiency in research and advanced algorithm development and the strength of the commercial AI ecosphere. Based on these factors, Ding derived what he calls an "AI Potential Index" and found that China rated only 17 versus 33 for the United States.[15] Ding points out, for example, that only about four percent of AI patents initiated in China are later also filed in other jurisdictions, a likely indicator of low quality. In testimony before a U.S. congressional committee in June 2019, he argued that China's purported rise to AI dominance has been overhyped, that the U.S. continues to have significant structural advantages and that American policy should be focused on maintaining the status quo.[16]

Kai-Fu Lee, in contrast, believes that the United States will likely continue to have an edge in research at the very frontier of artificial intelligence, but that this advantage will soon be overwhelmed by Chinese proficiency in doing the practical, nuts-and-bolts work of actually implementing the technology in applications across the economy. Lee argues that putting AI to work in the commercial sphere doesn't require top-flight visionary researchers but simply large numbers of competent and diligent engineers with easy access to a deluge of data that can be used to train machine learning algorithms.[17]

The stakes of any perceived AI race between the United States and China are raised greatly by the obvious reality that the impact of artificial intelligence will by no means be limited to the commercial sector. AI will deliver massive advantages that can be widely leveraged in military and national security applications. The Chinese government is keenly aware of this and has moved aggressively to erase any line between these two spheres. In 2017, in response to a direct initiative by Xi Jinping, the Chinese constitution was modified to explicitly require that any technological advances generated in the commercial sector must be shared with the People's Liberation Army. This is known as the principle of "military-civil fusion." In 2018, Baidu partnered with a Chinese military institute focused on electronic warfare technology on a project to develop intelligent command and control technology for the military. The Baidu executive in charge, Yin Shiming, was yet another example of an engineer who had developed deep experience working at Western companies, including SAP and Apple. At an event announcing the partnership, Yin declared that Baidu and the military institute would "work hand in hand to link up computing, data and logic resources to further advance the application of new generation AI technologies in the area of defense."[18]

Contrast this with the pressure that unhappy employees put on Google to end its bid to compete for the Pentagon's JEDI cloud computing contract. Another defense initiative, Project Maven, which involved the development of computer vision algorithms that could be used to analyze images collected from U.S. military drones, generated even more outrage among Google workers. In 2018, more than 3,000 employees signed a petition objecting to the project, and a number of technical experts left the company.[19] As with JEDI, Google ultimately abandoned the project. Though Google employees certainly have every right to express their views, the asymmetry here is, I think, both obvious and disturbing. The idea that workers at Baidu or Tencent could (or would) lodge a comparable protest is, frankly, absurd. There's no getting around the fact that the freedoms enjoyed by the citizens of democratic countries are not intrinsic human rights that exist simply to be exercised—rather, they are political rights that have to be defended in the face of authoritarianism. As overall proficiency in artificial intelligence technologies approaches parity between the two countries, how can the United States compete on a national security footing if companies like Google are reluctant to cooperate with American military and security agencies, while their Chinese counterparts face an obligation to assist China's authoritarian regime so explicit that it is written into the country's constitution?

To me, it seems clear that the United States and other Western countries need to take China's rapid rise in artificial intelligence very seriously. This will likely call for increased government support for basic research at universities. It's also critical that the U.S., in particular, continues to leverage one of its most important advantages: the fact that its universities and technology companies have been a magnet for talent from around the globe. The need for U.S. openness to high-skill immigration is clearly demonstrated by the backgrounds of the twenty-three top AI researchers I interviewed for my 2018 book *Architects of Intelligence*. Nineteen of the twenty-three people I spoke with currently work in the U.S. Of those nineteen, however, more than half were born outside the United States. Countries of origin include Australia, China, Egypt, France, Israel, Rhodesia (now Zimbabwe), Romania, and the United Kingdom. If the United States fails to continue attracting the brightest computer scientists from around the world, China will inevitably gain an advantage as it continues to invest more in educating a population that is roughly four times that of the United States.

THE RISE OF CHINA'S SURVEILLANCE STATE

NOWHERE IS THE powerful synergy between China's authoritarian system of government and its entrepreneurial artificial intelligence ecosphere more evident than in the exploding cluster of startup companies focused on facial recognition technology. As of early 2020, four companies in this group—SenseTime, CloudWalk, Megvii and Yitu—had achieved "unicorn" status, or a market valuation of more than a billion dollars.[20] While analysts may debate whether China has reached something close to parity with the United States in the overall technology of artificial intelligence, there is little doubt that when it comes to deep learning algorithms deployed to analyze and recognize human faces and other attributes, Chinese companies are at the absolute forefront of the field. As with other areas where artificial intelligence is being deployed in China, a critical driver of all this progress is access to a massive deluge of data that can be used to train machine learning algorithms. With an estimated 300 million surveillance cameras installed throughout the country as of 2020, China, when it comes to the availability of digital photographs of human faces in every conceivable situation and from every possible angle, is far and away the global leader.

Facial recognition startups are buoyed by seemingly limitless demand for surveillance technology from every level of China's authoritarian state. Some of the most eager purchasers of the technology are local police departments, who are increasingly setting up oppressive surveillance networks specific to their regions. While Xinjiang remains ground zero for the Chinese surveillance state, the technologies tested and perfected there are rapidly spreading across the country. Police departments often combine facial recognition systems with other technologies such as mobile phone scanners, which capture a unique identification code for every phone that passes through the vicinity, car license plate readers and fingerprint recognition technology to weave an Orwellian tapestry that is becoming increasing integrated over time. Algorithms can often, for example, match phone identification codes with faces, creating a comprehensive tracking and identification system for individuals. Such systems are installed in neighborhoods or at the entrances of specific buildings known to be associated with higher crime levels. Entry to housing complexes is also often enabled by facial recognition systems rather than by key cards or other less intrusive methods. This enables building management and local police departments to track residents and guests and also prevent illegal subletting of apartments.[21]

Surveillance cameras are also heavily clustered in any areas visited by travelers or where crowds are likely to congregate, such as train stations, stadiums, tourist attractions and event venues. In a number of highly publicized cases, police have arrested specific individuals at concerts or festivals attended by as many as 60,000 people, purely on the basis of algorithms alerting authorities to a facial recognition match.[22] In a scene seemingly straight out of a dystopian science fiction movie, police can apprehend suspects by wearing experimental facial recognition glasses that can generate an identification as long as the target remains still for several seconds and is included in a regional facial recognition database. Other AI systems can track people based on the clothing they are wearing or even through analyzing the unique characteristics of their gait.

In one of the most famous uses of this technology, the city of Xiangyang set up a system to catch and then embarrass jaywalkers at a busy intersection. The system captures photos of people crossing the street illegally that are then later matched with their identities and displayed on a large screen in an attempt to subject them to public shaming and gossip.[23] In other cities, including Shanghai, similar systems issue fines. To be sure, not all uses of facial recognition technology in China are specifically geared toward surveillance. The country is a leader in scanning faces to authorize payments at retail stores, to purchase train tickets or to board aircraft. However, any of the data generated through routine use of the technology is almost certainly available to police departments and security agencies.

While much of the pervasive surveillance in China can, at least to some extent, be defended as a mechanism to protect society from individuals with known criminal backgrounds, in other cases it violently transgresses ethical boundaries in ways that would be unthinkable in the West. Some police departments, for example, have put out specific requests for technology configured not to recognize individual faces, but rather the racial characteristics of Uyghurs or other "sensitive peoples."

Chinese facial recognition startups have moved rapidly to supply market demand. An April 2019 article by Paul Mozer of the *New York Times* included a screenshot of online marketing material from CloudWalk that promised potential buyers of its technology that if "the number of sensitive groups of people in the neighborhood increases (for example, if originally one Uyghur lives in a neighborhood, and within 20 days six Uyghur appear), it immediately sends alarms so that law enforcement personnel can respond, question the people and handle the situation, and develop a contingency plan."[24]

Next to images that appear to show a Uyghur family standing peacefully, a line of Uyghurs filing past militarized police and scenes of civil unrest, CloudWalk's website goes on to explain under the heading "Neighborhood control and prevention of sensitive peoples" that "in the neighborhood, the facial recognition system collects these people's identity and facial data, at the same time the Fire Eye big data platform collects sensitive groups' identities, times of entry and exit, the number of individuals, etc., and issues warnings to police so they can carry out their goal of managing and controlling sensitive groups."[25]

Even the branding of CloudWalk's product "Fire Eye big data platform" seems stripped directly from an especially frightening science fiction novel. The total lack of any attempt at subterfuge, or subtlety, in describing the technology's intent, even on a publicly available corporate website, is a pretty dramatic indication of just how overtly oppressive the Chinese government's campaign against the Uyghurs is, and how artificial intelligence, in the wrong set of hands, can be leveraged in genuinely dystopian ways. The danger here is by no means limited to China. Nearly any advanced facial recognition technology could conceivably be weaponized against specific groups by configuring the system to identify attributes such as race, gender, facial hair or religious attire.

China's trajectory toward ever more comprehensive surveillance of its citizens may culminate with the full implementation of the country's planned social credit system. Announced in 2014 as a way to reward "trustworthiness" throughout the population, the program's declared intent is to "allow the trustworthy to roam everywhere under heaven while making it hard for the discredited to take a single step."[26] The social credit system begins with measures that are typical of commercially administered credit or consumer rating systems in the West, such as evaluations based on a person's history of paying debt obligations or the kind of rating systems used on services like Uber or Airbnb. But the Chinese system goes much further, potentially intruding into virtually every aspect of daily life by taking into account violations of the law, as well as any behaviors deemed undesirable by the state. In addition to failing to pay your bills or fines in a timely manner, this might include playing too many video games, posting controversial thoughts on social media, associating with the wrong people, eating, littering or playing loud music on public transit, smoking where it is prohibited or even failing to properly sort garbage.[27] The social credit calculation can also reward positive behaviors, such as winning a civic or employee award, giving money to charity or making an outsized effort to take care of family members or assist neighbors. The system can reach even into the most intimate consumer decisions, for example, rewarding purchases deemed positive, such as baby diapers, while penalizing excessive purchases of alcohol. Those who achieve superior scores are rewarded with perks such as discounts on heating bills, shorter wait times at hospitals or government agencies, or preferred access to the best employment opportunities. Those who, on the other hand, have low social credit scores face penalties such as an inability to book tickets on planes and trains, lack of access to the best schools for their children or blocks on attempts to make reservations at desirable hotels or resorts. Once such a comprehensive system is fully operational, it will be an extraordinary intrusive mechanism of control, exerted continuously on virtually every adult in China's massive population. It's an idea that Human Rights Watch has appropriately called "chilling."[28]

While all that represents the ultimate vision, the current reality is far less cohesive. In practice, the social credit system is fragmented into experimental programs run by various city and local governments together with an array of commercial rating systems administered by corporations, like Alibaba or Tencent, that maintain mobile payment systems.[29] Some of the programs, such as one in the city of Rongcheng, have met with widespread approval from the public because they are relatively transparent, penalize only behaviors that are clearly illegal and have produced undeniably positive results. Drivers in Rongcheng, for example, began stopping at crosswalks in deference to pedestrians once it was made clear that violations would negatively impact their social credit ratings. Millions of airline or high-speed rail tickets have indeed been denied to Chinese citizens, but this is generally because their names have appeared on long-used blacklists rather than the result of an algorithmically generated score. The most important blacklist, maintained by the Supreme People's Court, primarily includes people who have unpaid debts, court judgments or fines—although, as with nearly all government functions in China, corruption and a lack of transparency are a constant problem. Over time, it seems inevitable that these systems will become far more integrated, and their intrusiveness will be amplified by facial recognition and other AI technologies used to track and monitor citizens. Eventually, a genuinely Orwellian system of comprehensive and carefully orchestrated social control could well emerge.

None of this is limited to China itself. Indeed, the export of surveillance technology plays a key role in the Chinese government's overall strategy to transform the country's production away from low-margin commodities and toward higher-value technology products. China controls nearly half of the global market for facial recognition technology. Much of this is led by a single Chinese firm, the telecommunications company Huawei. According to a September 2019 analysis by the Carnegie Endowment for International Peace, Huawei has sold surveillance technology, including facial recognition, to at least fifty countries and 230 cities, far more than any other single company. By comparison, the closest U.S. competitors—IBM, Palantir and Cisco—have each installed systems in less than a dozen countries.[30] Nations with authoritarian governments, such as Saudi Arabia and the United Arab Emirates, are especially eager customers for Chinese technology as they expand their own countrywide surveillance systems. In these nations, facial recognition is often already a routine aspect of daily life. I learned this myself on a trip to Abu Dhabi in early 2019 where I heard a widely circulated story about a wealthy woman who lost an expensive ring. She reported the mishap to authorities, who immediately applied facial recognition software to surveillance footage of the relevant area—and arrived at the doorstep of the person who had picked up the ring within hours of the incident.

Huawei's sales of surveillance equipment are often financed by loans backed by the Chinese government. Countries including Kenya, Laos, Mongolia, Uganda, Uzbekistan and Zimbabwe have participated, in some cases as part of Beijing's global Belt and Road Initiative, which is funding infrastructure in nearly seventy countries. Africa is an increasingly important focus, and by some accounts, Chinese facial recognition systems have already had a significant impact. Huawei claims, for example, that the installation of its technology in and around Kenya's capital city of Nairobi led to a forty-six percent reduction in crime in 2015.[31]

The security and human rights implications of the technologies being developed by Chinese companies have already led to significant friction with the United States. In May 2019, Huawei was subjected to trade restrictions that resulted in a ban on the sale of U.S. technology such as software and computer chips to the company. This was driven by some combination of U.S. posturing amidst an escalating general trade war and long-held concerns that the 5G mobile phone infrastructure technology sold by the company could potentially allow the Chinese government to access communications in the United States if the equipment were installed locally.[32] The U.S. also had mixed success in its effort to put significant pressure on allied countries to likewise prohibit use of Huawei's equipment. In addition, Huawei was accused of violating the U.S. trade embargo against Iran and had been singled out for having received inappropriate support from the Chinese state.

Five months later, the United States extended the trade blacklist to include several of China's most important artificial intelligence startups, as well as twenty Chinese police departments or security agencies, ostensibly because of human rights violations resulting from the deployment of their technology against the Uyghurs and other minority populations. Included in the ban were three of China's four facial recognition unicorns, as well as iFlytek, which specializes in speech recognition systems, and two companies that manufacture cameras and other surveillance hardware.[33]

In the wake of the coronavirus pandemic, tensions between the United States and China have escalated significantly, and there is widespread recognition that overdependence on Chinese production potentially threatens U.S. access to critically important strategic materials, as well as healthcare supplies and medicines. Even before the crisis, it was clear that the economic synergy and interdependence between the two countries—a phenomenon that the historian Niall Ferguson dubbed "Chimerica" in 2006—was gradually unwinding. If tensions increase further and the two countries continue to decouple, it seems inevitable that conflict and competition centered on the development and deployment of artificial intelligence will play a central role, and as it becomes increasingly evident that AI is both a systemic and a strategic technology, the specter of a full-out AI arms race between the two countries looms as a genuine danger.

AN EMERGING DEBATE OVER
FACIAL RECOGNITION IN THE WEST

In February 2019, the Indiana State Police were investigating a crime that had occurred when two men got into a fight at a park. One of the men pulled out a gun, shot the other man in the abdomen and then fled the scene. A bystander had recorded the incident on a mobile phone, so the state police detectives decided to try uploading an image of the assailant's face to a new facial recognition system with which they had been experimenting. The system generated a match immediately; the shooter had appeared in a video posted on social media along with a description that included his name. All told, it took about twenty minutes to solve the case, in spite of the fact that the suspect had not been arrested previously and did not even have a driver's license.[34]

The detectives interfaced with the facial recognition system through a mobile app provided by a mysterious company called Clearview AI. The database of photographs available to the Clearview app was truly massive. Rather than relying on official government photographs such as those associated with passports, driver's licenses or mug shots, the company had simply scoured the internet and scraped publicly available images from a variety of sources including Facebook, YouTube and Twitter.

If the Clearview system found a match, the app displayed links
to the webpages or social media profiles where a matching
photo had appeared online—often allowing for an immediate
identification. The dataset constructed by Clearview included
roughly three billion scraped images—more than seven times
the size of the official photographic database of U.S. citizens
maintained by the FBI. This was a remarkable feat, especially
since Clearview AI was a tiny company—orders of magnitude
smaller than the facial recognition unicorns in China—and, at
least until January 2020, was almost entirely unknown outside
of law enforcement circles.[35]

That month the *New York Times* published a major investi-
gative exposé by technology reporter Kashmir Hill that delved
into the company's background and, for the first time, shined a
spotlight on its operations. It turned out that Clearview, which
listed a nonexistent New York address on its LinkedIn page,
had been founded in 2016 by an Australian serial entrepreneur
named Hoan Ton-That. Among other funding, the startup had
received $200,000 in seed money from Silicon Valley venture
capitalist Peter Thiel, who had also co-founded Palantir, a data
analytics and surveillance company with significant ties to security agencies and police departments.

Clearview claimed that it made its technology available only
to legitimate law enforcement or government security agencies.
However, there was nothing in theory to prevent the company from eventually making its system available to the public, raising the specter of a near complete loss of anonymity.
Once the technology became widely available, virtually anyone,
anywhere, could be instantly identified by a random stranger
wielding Clearview's app. Given a person's name, it would be
a simple matter to find a home address, place of employment
and all kinds of other sensitive information. The inevitable outcome would be an explosion of stalking, blackmail or public

shaming for virtually any indiscretion, and countless other misbehaviors. In other words, it seems quite possible that an overreaching surveillance dystopia—potentially more intrusive and frightening than anything being contemplated in China—could emerge from the private sector in the United States, without any government involvement or oversight at all. Some backers of Clearview AI didn't seem especially concerned about this possibility. "I've come to the conclusion that because information constantly increases, there's never going to be privacy," one of the company's early investors told the *New York Times*. "Laws have to determine what's legal, but you can't ban technology. Sure, that might lead to a dystopian future or something, but you can't ban it."[36]

The *Times* article unleashed a storm of controversy focused on the company and also attracted the attention of hackers, who managed to break into Clearview's servers and obtain a complete list of the company's paying clients as well as prospective customers who were using a thirty-day free trial version of the app. It turned out that Clearview's users included major agencies such as the FBI, Interpol, U.S. Immigration and Customs Enforcement (ICE) and the U.S. Attorney's Office for the Southern District of New York, as well as hundreds of police departments across the globe. And despite the company's claims that it worked only with credentialed law enforcement agencies, the app was being used at private companies including Best Buy, Macy's, Rite Aid and Walmart. Worse still, there was evidence that private sector workers might be using the app without authorization from their employers. An investigation by *BuzzFeed* found that five accounts associated with Home Depot had performed nearly one hundred searches using the app, despite the fact that Home Depot's management claimed to be completely unaware of this.[37] In other words, access to the technology was already seeping out into the broader public sphere.

The publicity brought about an immediate backlash. Within weeks, Twitter, Facebook and Google had sent cease-and-desist orders to the company demanding that it stop scraping photographs from their servers and immediately delete any images already in its database.[38] By the end of February, Apple had disabled Clearview's iPhone app because the company had violated Apple's service agreement by circumventing the App Store.[39] Shortly thereafter, the company announced it would terminate all licensing agreements with private companies and focus exclusively on enforcement agencies, but this was widely dismissed as insufficient. In May, the American Civil Liberties Union filed a lawsuit against Clearview and declared that the company's technology posed "a nightmare scenario" that would "end privacy as we know it if it isn't stopped."[40] Clearview continues to operate and has stated that it believes it has the right to scour the internet for photographs and is prepared to do legal battle with the social media companies over such access.

Clearview AI offers an important cautionary tale not just for facial recognition but for artificial intelligence more generally. Wielding such a uniquely powerful technology, the smallest team of technical experts—or perhaps even a single individual—could conceivably unleash social or economic disruption on a nearly unimaginable scale, and as we'll see in the next chapter, the risks are by no means confined to the deployment of AI-enabled surveillance technologies.

Given the overwhelming backlash against the company, Clearview's ambitions seem likely to be reined in. More generally, however, the deployment of facial recognition is accelerating rapidly throughout Western countries, and democratic societies will face an increasingly urgent need to make value-based trade-offs and confront the ethical issues that surround the use of the technology. In London, which is already by far the most surveilled Western city, with more CCTV cameras per capita then Beijing,[41] facial recognition systems began to be deployed in early 2020. The Metropolitan Police stated that the systems would seek only people on a "bespoke" watch list consisting of individuals wanted for serious or violent crimes. However, the system could also be used to search for missing children and adults.[42]

In the United States, about a quarter of police departments have access to facial recognition technology. The systems are also widely employed at airports, where they search for known terrorists or criminals and are increasingly used to verify identity during the security screening process. In most cases, as with the system in London, the technology is used to identify only individuals on specific watch lists. However, we are gradually inching closer to the possibility of the kind of overreaching dystopia heralded by Clearview's app, where virtually anyone can be identified. The FBI maintains photographic databases that, according to a 2016 analysis by the Georgetown University Law School's Center on Privacy and Technology, include images for about 117 million people—roughly half the adult population of the United States.[43] Many of the images come from driver's license photographs maintained by the states and include all residents with identification cards issued by the state, not just those who are wanted for crimes or have criminal records. There is, needless to say, no requirement that individuals give consent for their photographs to be included, and no way to opt out from the system.

Though the potential threat to privacy is very real, it's important to recognize that, properly and ethically deployed, facial recognition systems do bring unambiguous benefits. Many dangerous criminals have been apprehended using the technology. While in the case of Clearview, I would say the privacy issues pretty clearly outweigh any advantages, the app nonetheless did lead to the apprehension of dangerous criminals and proved to be particularly effective at identifying sexual predators and purveyors of child pornography. Facial recognition systems deployed in public places likewise can offer real benefits in terms of lower crime rates. London's Metropolitan Police are not wrong when they say, "We all want to live and work in a city which is safe: the public rightly expect us to use widely available technology to stop criminals."[44]

Indeed, the widespread surveillance being deployed in China, while clearly oppressive from a Western perspective, is not necessarily viewed in a negative way by most of the Chinese public. Many residents of Xiangyang are very supportive of the jaywalking system because it worked, and the once-dangerous intersection is now more orderly. I have personally talked to a number of people who live in China, and one observation that comes up again and again is an increased sense of safety from crime and, in particular, peace of mind for parents of young children. The potential importance of this should not be underestimated. A sense of security in one's neighborhood is valued highly by most people and is correlated with better physical and mental health. This is one area where China, in many cases, is arguably outperforming the United States.

A safe environment is especially important for children. Jonathan Haidt, an author and professor at New York University, has been a strong advocate for "free range" parenting. Haidt argues that, in the United States, we have created a culture that is becoming dangerously overprotective of children and likely robs them of important opportunities to have the kind of unsupervised experiences that will help them develop into confident adults.[45] For most American parents, the idea of allowing young children to walk to school or play in a neighborhood park without oversight is a terrifying, and in some places illegal, prospect. I suspect that young children in China are not especially aware of the overreaching Orwellian state.

They do know, however, that they can walk to school or play in the park. It would be extraordinarily ironic if China's oppressive system of surveillance turns out to have a silver lining, at least where its youngest citizens are concerned. In time, that might actually help produce a more adventurous and innovative generation of young people. No one wants China's system in the United States, but to the extent that AI-based surveillance technologies can drive down crime rates and create safer environments, the trade-offs should be given careful consideration.

Though facial recognition can bring real benefits to society, it's critical that the technology be applied fairly and that the impact is equitable across demographic groups. And, here, there is a major problem. In a number of studies, facial recognition systems have consistently been shown to exhibit some degree of both racial and gender bias. To be sure, this is nothing like the Chinese algorithms explicitly designed to seek out Uyghurs, but rather results from a preponderance of white male faces in the datasets used to train deep learning algorithms. The faces in one commonly used training dataset are eighty-three percent white and seventy-seven percent male.[46] The problem generally manifests as an increased likelihood of a "false positive" result for non-white and female faces. In other words, women and people of color are more likely to generate an incorrect match.

In 2018, the American Civil Liberties Union compared images of all 538 members of the U.S. Congress to a large dataset of booking photographs taken of people who had been arrested. The ACLU used the Rekognition system available through Amazon Web Services, which is becoming increasingly popular with police departments because it is available at a very low cost. The ACLU was able to conduct the experiment for only about $12. The system flagged twenty-eight members of Congress as arrestees whose photos were contained in the mug shot dataset. Assuming that none of the individuals who were arrested have, in fact, been elected to either the House or the Senate, these were all false positives. Aside from the sheer number of errors, a major concern was that the false positives generated by the system were weighted heavily toward non-white members of Congress. People of color represent about twenty percent of Congress, but they accounted for thirty-nine percent of the incorrect matches. In response to the study, Amazon argued that the ACLU had configured the system incorrectly because it used the default confidence threshold of eighty percent for matches, rather than a more appropriate ninety-five percent. The ACLU, however, noted that Amazon provides no specific instructions regarding the proper setting, and that many police departments would likely leave the system set to its defaults.[47]

A much more comprehensive study was performed in 2019 by the National Institute of Standards and Technology (NIST), which is part of the U.S. Department of Commerce. NIST evaluated 189 facial recognition systems from ninety-nine different companies.[48] They found that, in nearly all cases, false positives were lowest for European faces and significantly higher for African and Asian faces. The predictable exception was for algorithms developed by Chinese companies, for which East Asian faces produced the most accurate results. The systems were also generally more accurate for male than for female faces, although the magnitude of the difference was smaller than for different races.

The difference in accuracy for non-white races was substantial. A black person, for example, might face a likelihood of a false positive more than a hundred times that of a white person. In other words, an African American might be a hundred times more likely than a white person to be incorrectly flagged as a potential offender and thereby inconvenienced, accosted or perhaps even detained. Essentially, this amounts to a digital rendition of the real-world scenarios already familiar to African Americans, where, for example, they are often followed by security staff in retail stores or given undue attention by sales clerks.

In theory, it should be possible to address this problem simply by including more diverse faces in the training datasets. However, the companies that develop facial recognition systems have often struggled to find high-quality images of non-white faces obtained ethically and with consent—or in other words, without resorting to techniques like scraping images from the internet, as was done by Clearview.[49] Solutions to this problem can sometimes raise questions of their own, and this is an area where companies willing to push ethical boundaries may sometimes obtain an advantage. In 2018, the Chinese unicorn CloudWalk entered into a controversial agreement with the government of Zimbabwe to build a comprehensive facial recognition system for the country. As part of the agreement, CloudWalk will obtain access to photographs of Zimbabwean citizens and will be able to use them to train its machine learning algorithms. The resulting systems could potentially be deployed anywhere in the world, without, of course, any knowledge or consent from the citizens of Zimbabwe.[50]

Issues like these, as well as the situation with Clearview, make it clear that facial recognition cannot be left in the hands of an unregulated private sector. Regulation and oversight of the technology is essential. Had the *New York Times* not exposed the company, Clearview's technology might have reached the public without any oversight whatsoever, long before there was any general awareness of the threat to privacy that it represented. At a minimum, we need clear regulations to ensure the fairness of any algorithms deployed, as well as safeguards that will prevent deployment of surveillance systems in ways that threaten public privacy.

In the absence of general standards, some jurisdictions, such as San Francisco, have already taken the initiative to completely

ban the use of facial recognition technology by police departments and local government. However, this does not extend to
the private sector. As in China, it's becoming increasingly common for facial recognition to be deployed as an entry mechanism for large housing developments. Some residents have filed
lawsuits arguing that this represents an invasion of their privacy.
Retail stores likewise can deploy the technology with few con-
straints. Clearly, we need regulation at the national level that
defines a basic set of rules that must apply to systems deployed
either publicly or privately. Attitudes toward privacy, surveillance and the importance of public safety vary, and it seems
very likely that individual countries, regions and cities will make
different trade-offs around the risk-value proposition of facial
recognition and other AI-based surveillance technologies. In
democratic societies, there should be a transparent process that
incorporates public input, and the technology needs to be governed by a set of foundational principles that protect the rights
of all involved.

THE VERY REAL possibility of an AI arms race with China, unprecedented threats to personal privacy and new forms of discrimination are just a few of the emerging dangers as the technology
of artificial intelligence continues to relentlessly advance. In the
next chapter, we'll take a broader view of some of the risks that
are inherently coupled with AI and discuss which dangers need
our immediate attention and which are more speculative concerns likely to arise only in the far future.

CHAPTER 8

AI RISKS

IT'S EARLY NOVEMBER, JUST TWO DAYS BEFORE THE PRESIDENTIAL election in the United States. The Democratic candidate has spent much of her career fighting to advance civil rights and expand protections for marginalized communities. Her record on this issue is seemingly impeccable. It therefore comes as an immense shock when an audio recording that purports to be of the candidate engaging in a private conversation appears, and then immediately goes viral, on social media. In the conversation, the candidate not only uses explicitly racist language, she also openly admits, and even laughs about, her lifelong success in keeping her bigotry hidden.

Within an hour of the audio clip's appearance, the candidate vehemently denies its authenticity. No one who knows her personally believes that the words can possibly be hers, and dozens of people come forward to support her. Anyone who chooses to believe her, however, has to confront a very uncomfortable reality: *it is her voice.* Or at least to virtually any human ear, it appears to be the candidate speaking. The distinct way in which she enunciates certain words and phrases, the cadence of her speech, all seem to undeniably belong to the woman that most people expect will soon become the president-elect of the United States.

As the audio recording explodes on the internet and is played repeatedly on cable TV, the social media universe reels with confusion and outrage. Before securing the nomination, the candidate fought a vicious primary battle, and now some angry supporters of her opponents begin to call for her to step aside.

The campaign immediately hires a panel of experts to independently review the audio file. After a day of intense analysis, they declare that the recording is likely a "deepfake"—audio generated by machine learning algorithms that have been extensively trained on examples of the candidate speaking. There have been warnings about deepfakes for years, but so far they have been rudimentary and easy to identify as fabrications. This example is different; it is clear that the state of the technology has advanced significantly. Even the panel of experts cannot state with absolute certainty that the audio file is a fake and not an actual recording.

Based on the determination of the expert panel, the campaign succeeds in having most online copies of the audio file taken down. However, millions have already heard the words. As Election Day dawns, a number of critical questions loom: Has everyone who heard the recording learned that it is likely a fake? Can voters who are told the recording is a fabrication manage to somehow "unhear" the hateful words that are by now indelibly etched into their memories—especially if they happen to belong to a group targeted in the conversation? Will the audio clip depress turnout within the communities that the Democratic candidate most relies on? And if she loses, will a majority of the American people feel that the election has been stolen? What will happen then?

While the scenario above is obviously fiction, the reality is that something similar to what I have described could happen—

perhaps within just a few years. If you doubt this, consider that, in July 2019, the cybersecurity firm Symantec revealed that three unnamed corporations had already been bilked out of millions of dollars by criminals using audio deepfakes.[1] In all three cases, the criminals did this by using an AI-generated audio clip of the company CEO's voice to fabricate a phone call ordering financial staff to move money to an illicit bank account. CEOs—like the presidential candidate imagined above—typically have a rich trove of online audio data (speeches, television appearances, etc.) that could be used to train machine learning algorithms. Because the technology is not yet at the point where it can produce truly high-quality audio, the criminals in these cases intentionally inserted background noise (such as traffic) to mask the imperfections. However, the quality of deepfakes is certain to get dramatically better in the coming years, and eventually, things will likely reach a point where truth is virtually indistinguishable from fiction.

Malicious deployment of deepfakes, which can be used to generate not only audio but also photographs, video and even coherent text, is just one of the important risks we face as artificial intelligence advances. In the previous chapter, we saw how AI-enabled surveillance and facial recognition technologies could destroy the very concept of personal privacy and lead us into an Orwellian future. In this chapter, we will look at some of the other major concerns that are likely to arise as AI becomes ever more powerful.

WHAT IS REAL, AND WHAT IS ILLUSION? DEEPFAKES AND THREATS TO SECURITY

Deepfakes are often powered by an innovation in deep learning known as a "generative adversarial network," or GAN. GANs deploy two competing neural networks in a kind of game that relentlessly drives the system to produce ever higher quality simulated media. For example, a GAN designed to produce fake photographs would include two integrated deep neural networks. The first network, called the "generator," produces fabricated images. The second network, which is trained on a dataset consisting of real photographs, is called the "discriminator." The images synthesized by the generator are mixed with actual photographs and fed to the discriminator. The two networks interact continuously, engaging in a contest in which the discriminator evaluates each photograph produced by the generator and decides whether it is real or fake. The generator's objective is to try to deceive the discriminator by passing off faux photographs. As the two networks continue their iterative battle, the image quality gets better and better until ultimately the system reaches a kind of equilibrium in which the discriminator can do no better than guess at the authenticity of the images it analyzes. This technique produces astonishingly impressive fabricated images. Search the internet for "GAN fake faces," and you'll find numerous examples of high-resolution images that portray completely nonexistent individuals. Try stepping into the role of the discriminator network. The photographs seem completely real, but they are an illusion—a rendering conjured up from the digital ether.

Generative adversarial networks were invented by a graduate student at the University of Montreal named Ian Goodfellow. One evening in 2014, Goodfellow went out to a local bar with a few of his friends. They discussed the problem of building a deep learning system that could generate high-quality images. After imbibing an unknown number of beers, Goodfellow proposed the basic concept behind a generative adversarial network, but was met with extreme skepticism. Afterward, Goodfellow went home and immediately began coding. Within hours, he had the first functioning GAN. The accomplishment would turn Goodfellow into a legendary figure within the deep learning community. Yann LeCun, the chief AI scientist at Facebook, says generative adversarial networks are "the coolest idea in deep learning in the last 20 years."[2] After completing his PhD at the University of Montreal, Goodfellow went on to work at the Google Brain project and OpenAI and is now a director of machine learning at Apple. He is also the primary author of the leading university textbook on deep learning.

Generative adversarial networks can be deployed in many positive ways. In particular, synthesized images or other media can be used as training data for other machine learning systems. For example, images created with a GAN might be used to train the deep neural networks used in self-driving cars. There have also been proposals to use synthetic non-white faces to train facial recognition systems as a way of overcoming the racial bias problem when sufficient numbers of high-quality images of real people of color cannot be ethically obtained. When applied to voice synthesis, GANs can be used to provide people who have lost the ability to speak with a computer-generated replacement that sounds like their own voice. The late Stephen Hawking, who lost his voice to the neurodegenerative disease ALS, or Lou Gehrig's disease, famously spoke in a distinctive computer synthesized voice. More recently, ALS patients like the NFL player Tim Shaw have had their natural voices restored by training deep learning systems on recordings made before the illness struck.

However, the potential for malicious use of the technology is inescapable and, evidence already suggests for many tech savvy individuals, irresistible. Widely available deepfake videos created with humorous or educational intent demonstrate what is possible. You can find numerous fake videos featuring high-profile individuals like Mark Zuckerberg saying things they would presumably never say—at least in public. One of the most famous examples was created by the actor and comedian Jordan Peele, who is known for his impersonation of Barak Obama's voice, in collaboration with *BuzzFeed*. In Peele's public service video intended to make the public aware of the looming threat from deepfakes, Obama says things like "President Trump is a total and complete dipshit."[3] In this instance, the voice is Peele's imitation of Obama, and the technique used alters an existing video by manipulating President Obama's lips so they synchronize with Peele's speech. Eventually, we will likely see videos like this in which the voice is also a deepfake fabrication.

An especially common deepfake technique enables the digital transfer of one person's face to a real video of another person. According to the startup company Sensity (formerly Deeptrace), which offers tools for detecting deepfakes, there were at least 15,000 deepfake fabrications posted online in 2019, and this represented an eighty-four percent increase over the prior year.[4] Of these, a full ninety-six percent involve pornographic images or videos in which the face of a celebrity—nearly always a woman—is transplanted onto the body of a pornographic actor.[5] While celebrities like Taylor Swift and Scarlett Johansson have been the primary targets, this kind of digital abuse could eventually be targeted against virtually anyone, especially as the technology advances and the tools for making deepfakes become more available and easier to use.

As the quality of deepfakes relentlessly advances, the potential for fabricated audio or video media to be genuinely disruptive looms as a seemingly inevitable threat. As the fictional anecdote at the beginning of this chapter illustrates, a sufficiently credible deepfake could quite literally shift the arc of history—and the means to create such fabrications might soon be in the hands of political operatives, foreign governments or even mischievous teenagers. And it's not just politicians and celebrities who need to worry. In the age of viral videos, social media shaming and "cancel culture," virtually anyone could be targeted and possibly have both their career and life destroyed by a deepfake. Because of its history of racial injustice, the United States may be especially vulnerable to orchestrated social and political disruption. We've seen how viral videos depicting police brutality can almost instantly lead to widespread protests and social unrest. It is by no means inconceivable that, at some point in the future, a video so inflammatory that it threatens to rend the very social fabric could be synthesized—perhaps by a foreign intelligence agency.

Beyond videos or sound clips intended to attack or disrupt, there will be nearly endless illicit opportunities for those who simply want to profit. Criminals will be eager to employ the technology for everything from financial and insurance fraud to stock market manipulation. A video of a corporate CEO making a false statement, or perhaps engaging in erratic behavior, would likely cause the company's stock to plunge. Deepfakes will also throw a wrench into the legal system. Fabricated media could be entered as evidence, and judges and juries may eventually live in a world where it is difficult, or perhaps impossible, to know whether what they see before their eyes is really true.

To be sure, there are smart people working on solutions. Sensity, for example, markets software that it claims can detect most deepfakes. However, as the technology advances, there will inevitably be an arms race—not unlike the one between those who create new computer viruses and the companies that sell software to protect against them—in which malicious actors will likely always have at least a small advantage. Ian Goodfellow says he doesn't think we will be able to know if an image is real or fake simply by "looking at the pixels."[6] Instead, we'll eventually have to rely on authentication mechanisms like cybernetic signatures for photos and videos. Perhaps someday every camera and mobile phone will inject a digital signature into every piece of media it records. One startup company, Truepic, already offers an app to provide this kind of capability. The company's

customers include major insurance companies that rely on photographs that their customers send in to document the value of
everything from buildings to jewelry and expensive trinkets.[7]
Still, Goodfellow thinks that ultimately there's probably not
going to be a foolproof technological solution to the deepfake
problem. Instead, we will have to somehow learn to navigate
within a new and unprecedented reality where what we see and
what we hear can always potentially be an illusion.

While deepfakes are intended to deceive human beings, a
related problem involves the malicious fabrication of data intended to trick or gain control of machine learning algorithms.
In these "adversarial attacks," specially designed inputs cause a
machine learning system to make an error in a way that allows
the attacker to produce a desired output. In the case of machine
vision, this involves placing something in the visual field that
distorts the neural network's interpretation of the image. In one
famous example, researchers took a photograph of a panda,
which a deep learning system identified correctly with about a
fifty-eight percent confidence level, and by adding carefully constructed visual noise to the image, they tricked the system into
being more than ninety-nine percent certain that the panda was
instead a gibbon.[8] An especially chilling demonstration found
that simply adding four small rectangular black and white stick-
ers to a stop sign tricked an image recognition system of the
type used in self-driving cars into believing the stop sign was instead a 45 mph speed limit sign.[9] In other words, an adversarial
attack could easily have life-or-death consequences. In both of
these cases, a human observer might not even notice—and certainly wouldn't be confused by—the information surreptitiously
added to the image. This is, I think, an especially vivid demonstration of just how shallow and brittle the understanding that
coalesces in today's deep neural networks really is.

Adversarial attacks are taken seriously within the AI research community and are viewed as a critical vulnerability. Indeed, Ian Goodfellow has devoted much of his research career to studying security issues within machine learning systems and developing potential safeguards. Building AI systems that are robust in the face of adversarial attacks is no easy task. One approach involves what is called "adversarial learning," or in other words intentionally including adversarial examples in the training data in the hope that the neural network will be able to identify attacks if they occur once the system is deployed. As with deepfakes, however, there is likely destined to be a perpetual arms race in which attackers will always have an advantage. As Goodfellow points out, "no one has yet designed a truly powerful defense algorithm that can resist a wide variety of adversarial example attack algorithms."[10]

Adversarial attacks are specific to machine learning systems, but they will become one more important item on the list of computer vulnerabilities that can be exploited by cybercriminals, hackers or foreign intelligence agencies. As artificial intelligence is increasingly deployed, and as the Internet of Things results in ever more interconnection between devices, machines and infrastructure, security issues will become vastly more consequential, and cyberattacks will almost certainly become more frequent. Wider deployment of AI will inevitably result in systems that are more autonomous, with fewer humans in the loop, and these systems will make increasingly attractive targets for cyberattack. Imagine, for example, that someday self-driving trucks deliver food, medicine and critical supplies. An attack that managed to bring these vehicles to a halt, or even create significant delays, could easily have life-threatening consequences.

The upshot of all this is that increased availability and reliance on artificial intelligence will come coupled with systemic security risk, and this will include threats to critical infrastructure and systems, as well as to the social order, our economy and our democratic institutions. Security risks are, I would argue, the single most important near-term danger associated with the rise of artificial intelligence. For this reason, it is critical that we invest in research focused on building AI systems that are robust, and that we form an effective coalition between government and the commercial sector to develop appropriate regulations and safeguards before critical vulnerabilities are introduced.

LETHAL AUTONOMOUS WEAPONS

Hundreds of miniature drones swarm through the U.S. Capitol building in a coordinated attack. Employing facial recognition technology, the drones identify specific individuals and then fly directly at them at high speed, carrying out targeted kamikaze assassinations by delivering a small shaped explosive that kills just as effectively as a bullet. The capitol is in complete chaos, but it later turns out that all the targeted members of Congress belonged to a single political party.

This is just one chilling scenario sketched out in the 2017 short film *Slaughterbots*.[11] Designed as a warning about the looming peril of lethal autonomous weapons, the video was produced by a team working with Stuart Russell, a professor of computer science at the University of California, Berkeley, who has focused much of his recent work on the inherent risks of artificial intelligence as the technology continues to advance. Russell believes that lethal autonomous weapons, which the United Nations defines as weapons that can "locate, select and eliminate human targets without human intervention,"[12] should be classified as a new type of weapon of mass destruction. In other words, these AI-powered weapons systems could ultimately be

as disruptive and destabilizing as chemical, biological or perhaps even nuclear arms.

   This argument is based primarily on the fact that once you
eliminate direct human control and authorization to kill, such
weapons become highly scalable in terms of the destruction they
can unleash. Any drone could potentially be used as a weapon,
and you could launch hundreds of them at a time, but if they are
controlled remotely you would also need hundreds of people to
pilot the devices. If the drones are fully autonomous, however, a
small team could deploy massive swarms, unleashing almost unimaginable carnage. As Russell told me, "Someone can launch
an attack, where five guys in a control room could launch
10,000,000 weapons and wipe out all males between the age of
12 and 60 in some country. So, these can be weapons of mass
destruction, and they have this property of scalability."[13] Given
the ability for facial recognition algorithms to discriminate on
the basis of ethnicity, gender or attire, it's easy to imagine truly
chilling scenarios involving automated ethnic cleansing, or mass
assassination of political opponents, carried out with a ruthless-
ness and speed that would have once been inconceivable.

   Even if we could completely set aside the truly dystopian
possibilities and assume the technology would be limited strictly
to legitimate military engagements, autonomous weapons raise
critical ethical concerns. Is it morally acceptable to give a machine the ability to independently take a human life, even if doing so might increase targeting efficiency and perhaps reduce
the risk of collateral damage to innocent bystanders? And in the
absence of direct human control, who should be held account-
able in the event of an error that results in injury or loss of life?

   The danger that the technology they're working to advance
might be deployed in such weapons stirs a great deal of passion
among many artificial intelligence researchers. More than 4,500

individuals as well as hundreds of companies, organizations and universities have signed open letters declaring their intention to never work on autonomous weapons and calling for a general ban on the technology. An initiative is underway within the United Nations Convention on Conventional Weapons to outlaw fully autonomous killing machines in much the same way that chemical and biological weapons are already forbidden. Progress, however, has been underwhelming. According to the Campaign to Stop Killer Robots, an advocacy group focused on a UN ban, as of 2019, twenty-nine mostly smaller or developing nations have formally called for a complete prohibition of autonomous weapons technology. The major military powers, however, are not on board. The one exception is China, which has signed on with the stipulation that it wants to ban only actual use of the weapons, allowing for their development and production.[14] The United States and Russia have both opposed a ban, and it therefore seems unlikely that the weapons will be completely outlawed anytime soon.[15]

My own view is rather pessimistic. It seems to me that the competitive dynamic and lack of trust between major countries will probably make at least the development of fully autonomous weapons a near certainty. Indeed, every branch of the U.S. military as well as nations including Russia, China, the United Kingdom and South Korea are actively developing drones with the ability to swarm.[16] Likewise, the U.S. Army is deploying armed robots that look like small tanks,[17] and the Air Force is reportedly developing an unmanned AI-driven fighter jet with the ability to defeat human-piloted aircraft in dogfights.[18] China, Russia, Israel and other countries are also deploying or developing similar technologies.[19]

So far, the United States and other major militaries have made a commitment that a human will always be kept in the loop, and that specific authorization will be required before these machines engage in an attack that could result in loss of life. The reality, however, is that full automation on the battlefield will deliver enormous tactical advantages. No human can possibly react and make decisions at a speed comparable to that of artificial intelligence. Once one country breaches the current informal prohibition on full autonomy and begins to deploy these capabilities, it's inevitable that any competing military would have to immediately follow suit or find itself at a critical disadvantage. This fear of falling behind is likely a major reason that the U.S., China and Russia are all opposed to a formal ban on the development and production of autonomous weapons systems.

I think we can get something of a preview of how all this might unfold by looking at another type of warfare—the continuous battle between AI-powered trading systems on Wall Street. Algorithmic trading now dominates the daily transactions on the major stock exchanges, accounting for as much as eighty percent of overall trading volume in the United States. As far back as 2013, a group of physicists studied financial markets and published a paper in the journal *Nature* declaring that "an emerging ecology of competitive machines featuring 'crowds' of predatory algorithms" existed and that algorithmic trading had perhaps already progressed beyond the control— and even comprehension—of the humans who designed the systems.[20] Those algorithms now incorporate the latest advances in AI, their influence on markets has increased dramatically, and the ways in which they interact have grown even more incomprehensible. Many algorithms, for example, have the ability to tap directly into machine-readable news sources provided by companies like Bloomberg and Reuters and then trade on that information in tiny fractions of a second. When it comes to short-term moment-by-moment trading, no human being can begin to comprehend the details of what is unfolding, let alone attempt to outsmart the algorithms. Eventually, I suspect the same will be true with many of the kinetic confrontations that occur on battlefields.

Even if autonomous battlefield technologies are deployed exclusively by major militaries, the dangers are very real. A robotic battle might unfold with a speed that could outstrip the ability of military or political leaders to fully understand or deescalate the situation. In other words, the risk of blundering into a major war as a result of a relatively minor incident might increase significantly. Another concern is that in a world where robots battle robots and very few human lives are immediately at stake, the perceived cost of going to war might become uncomfortably low. This is arguably already an issue in the United States, where the elimination of the draft in favor of an all-volunteer military has led to a situation in which very few of society's elites send their children to serve in the armed forces. As a result, those who hold the most power have very little skin in the game; they tend to be insulated from the direct personal costs of military action. I suspect that this disconnect may have contributed significantly to the U.S.'s decades-long engagements in the Middle East. To be sure, if a machine can go into harm's way and thereby preserve the life of a soldier, that is an unambiguously good thing. But we need to be quite careful that we don't allow that perception of low risk to color our collective judgment when it comes to making a decision to go to war.

The greatest danger is that legitimate governments and militaries might fail to maintain control of lethal autonomous technology once the weapons were produced. In that event, the weapons could end up being traded by the types of illicit arms dealers that deliver machine guns or other small armaments into the hands of terrorists, mercenaries or rogue states. If autonomous weapons were to become widely available, the nightmare scenarios depicted in the *Slaughterbots* video could easily become reality. And even if the weapons were not available for purchase, the barriers to developing the technology are far lower than for other weapons of mass destruction. In the case of drones especially, the same easy-to-obtain technology and components intended for commercial or recreational applications could potentially be weaponized. While building a nuclear weapon poses an enormous challenge even when the resources of a nation-state are available, designing and deploying a small swarm of autonomous drones is something that might be accomplished by a few people working in a basement. Much like a virus, once autonomous weapons technology escapes into the environment, it will be very difficult to defend against or contain, and chaos may well ensue.

One common mistake—sometimes abetted by the media—is to conflate the specter of lethal autonomous weapons with the science fiction scenarios we've all seen in movies like *The Terminator*. This is a category error—and a dangerous distraction from the near-term dangers posed by such weapons. The risk is not that the machines will somehow wrest themselves from our control and decide to attack us of their own volition. That would require artificial general intelligence, which as we have seen, likely lies at least decades in the future. Rather, we have to worry about what human beings will choose to do with weapons that are no more "intelligent" than an iPhone—but which are ruthlessly competent at identifying, tracking and killing targets. And this is by no means a purely futuristic concern. As Stuart Russell says in the conclusion of the *Slaughterbots* video, the film dramatizes "the results of integrating and miniaturizing technologies that we already have." In other words, these are weapons that could quite possibly emerge with the next few years, and if we want to prevent that, "the window to act is closing fast."[21] Given that an outright United Nations–sanctioned ban on the weapon may not be coming anytime soon, the international community should at a minimum focus on ensuring that such weapons never become accessible to terrorists or other non-state actors who might deploy them against civilians.

BIAS, FAIRNESS AND TRANSPARENCY IN MACHINE LEARNING ALGORITHMS

As artificial intelligence and machine learning are deployed more and more widely, it's critical that the results and recommendations produced by these algorithms are perceived as fair and that the reasoning behind them can be adequately explained. If you're using a deep learning system to maximize the energy efficiency of some industrial machine, then you are probably not particularly concerned about the details that drive an algorithmic outcome; you simply want the optimal result. But when machine learning is applied to areas like criminal justice, hiring decisions or the processing of home mortgage applications—in other words, to high-stakes decisions that directly impact the rights and future well-being of human beings—it's essential that algorithmic outcomes can be shown to be unbiased across demographic groups and that the analysis that led to those outcomes is transparent and just.

Bias is a common issue in machine learning, and in most cases the problem arises because of problems with the data used to train the algorithms. As we saw in the previous chapter, facial recognition algorithms developed in the West are often biased against people of color because the training dataset tends to contain overwhelmingly white faces. A more general problem is that much of the data that is used to train algorithms results directly from human behaviors, decisions and actions. If the humans who generate the data are biased in some way— for example on the basis of race or gender—then that bias will automatically be encapsulated in the training dataset.

As an example, consider a machine learning algorithm designed to screen resumes for an open job at a large corporation. Such a system might be trained on the full text of all the resumes received from past applicants to similar jobs, together with the decisions that hiring managers made for each of those resumes. The machine learning algorithm would churn through all this data and assimilate an understanding of the characteristics of a resume that will likely lead to a decision to bring a job candidate in for further interviews, as well as resume attributes that suggest the applicant should be rejected without further consideration. An algorithm that can do this effectively, generating a manageable list of top-ranked candidates, is likely to save a great deal of time when a human resource department has to weed through hundreds or thousands of potential applicants, and for this reason, resume screening systems like this are becoming popular, especially in large companies. Suppose, however, that the past hiring decisions upon which the algorithm is trained reflect some degree of overt or subconscious racism or sexism on the part of hiring managers. In that case, the machine learning system will automatically pick up that bias as it goes through its normal training process. There is no nefarious intent on the part of the creators of the algorithm; the bias exists in the training data. The result would be a system that perpetuated, or perhaps even amplified, existing human biases and would be demonstrably unfair to people of color or women.

Something very similar to this happened at Amazon in 2018 when the company halted development of a machine learning system because it was shown to be biased against women when screening resumes for technical positions. It turned out that when a resume included the word "women's," as might happen in a reference to women's clubs or sports or when the candidate has graduated from an all-women's college, the system gave the resume a lower score, putting the female job candidate at a disadvantage. Even when Amazon's developers made corrections for the specific problems it discovered, it wasn't possible to guarantee that the algorithm would be unbiased because other variables might be acting as a proxy for gender.[22] It's important to note that this did not necessarily imply outright sexism in prior hiring decisions. The algorithm may have been trained to be biased simply because women are underrepresented in technical roles, and therefore men constitute the vast majority of hires. According to Amazon, the algorithm never made it beyond the development phase and was never actually used to screen resumes, but if it had been deployed, it would unquestionably have worked to solidify the underrepresentation of women in technical jobs.

An even higher stakes situation occurs when machine learning systems are used in the criminal justice system. Such algorithms are often used to assist with making decisions on bail, parole or sentencing. Some of these systems are developed by state or local governments, while others are designed and sold by private companies. In May 2016, *Propublica* published an analysis of an algorithm called COMPAS that is widely used to predict the likelihood that a particular individual will be a repeat offender upon release.[23] The analysis suggested that African American defendants were unfairly being assigned a higher risk than white defendants. That assessment seemed to be supported by anecdotal evidence. *Propublica*'s article included the story of an eighteen-year-old black woman who rode a child's bicycle that was far too small for her for a short distance before abandoning it after the owner objected. In other words, something that seems more like an instance of mischievous behavior than a serious attempt at theft. The young woman was nonetheless arrested, and the COMPAS system was applied to her case when she was booked into jail to await a court appearance. It turned out that the algorithm assigned her a significantly higher risk of becoming a repeat offender than a forty-one-year-old white man who already had a prior conviction for armed burglary and had served five years in prison.[24] The company that sells the COMPAS system, Northpoint, Inc., disputes the analysis performed by *Propublica*, and there continues to be a debate about the extent to which the system is actually biased. It is especially concerning, however, that the company is unwilling to share the computational details of its algorithm because it considers them to be proprietary. In other words, there is no way for a third party to perform a detailed audit of the system for bias or accuracy. It seems clear that when algorithms are deployed to make decisions that are so extraordinarily consequential for human lives, there needs to be more transparency and oversight.

Though bias in the training data is the most common cause of unfairness in machine learning systems, it is not the only factor at play. The design of the algorithms themselves can also introduce or amplify bias. For example, suppose a facial recognition system was trained on a dataset that exactly mirrored the demographic distribution of the U.S. population. Because African Americans are only about thirteen percent of the population, the system could still end up biased against black people. The extent to which this became an issue—whether the problem was amplified or mitigated—would be determined by technical decisions made in the design of the algorithm.

The good news is that designing machine learning systems to be fair and transparent has become a major focus of AI research. All the major tech companies are making significant investments in this area. Google, Facebook, Microsoft and IBM have all released software tools designed to help developers build fairness into machine learning algorithms. Making deep learning systems explainable and transparent so that outcomes can be audited is a particular problem because deep neural networks tend to be a kind of "black box," in which analysis and comprehension of the input data is distributed across millions of connections between artificial neurons. Likewise, assessing and ensuring fairness is a very challenging and highly technical issue. As Amazon found with its resume screening system, simply tweaking the algorithm to ignore parameters like race or gender is not an adequate solution because the system might instead focus on proxies. For example, the first name of a job candidate might indicate gender, and the neighborhood or zip code could be a proxy for race. One especially promising approach to AI fairness is the use of counterfactuals. With this technique, a system is checked to verify that it produces the same outcome when sensitive variables like race, gender or sexual orientation are changed to different values. Still, research in these areas is just getting started, and it will take a lot more work to develop techniques that will consistently result in machine learning systems that are truly fair.

The ultimate promise of AI deployed in high-stakes decisions is a technology that reliably produces less bias and greater accuracy than human judgment alone. Though fixing bias in an algorithm can be challenging, it is nearly always much easier than doing the same for a human being. As McKinsey Global Institute Chairman James Manyika told me, "On the one hand, machine systems can help us overcome human bias and fallibility, and yet on the other hand, they could also introduce potentially larger issues of their own."[25] Minimizing or eliminating those fairness issues is one of the most critical and urgent challenges facing the field of artificial intelligence.

In order to achieve this outcome, it's also important that the developers building, testing and deploying AI algorithms come from diverse backgrounds. Given that artificial intelligence is poised to shape our economy and society, it is essential that the experts who best understand the technology—and are therefore best positioned to influence its direction—are representative of society as a whole. Progress in reaching this goal, however, has so far been limited. A 2018 study found that women represent only about twelve percent of leading artificial intelligence researchers, and the numbers for underrepresented minorities are even lower. As Stanford's Fei-Fei Li says, "If we look around, whether you're looking at AI groups in companies, AI professors in academia, AI PhD students or AI presenters at top AI conferences, no matter where you cut it: we lack diversity. We lack women, and we lack under-represented minorities."[26] The universities, the major tech companies and nearly all top AI researchers are firmly committed to changing this. One especially promising initiative was co-founded by Li: AI4ALL is an organization dedicated to attracting young women and underrepresented groups into the field of artificial intelligence by providing summer camps for talented high school students. The organization has expanded rapidly and now offers summer programs on eleven university campuses in the United States. While much work remains to be done, programs like AI4ALL together with an industry commitment to attracting inclusive AI talent will likely produce a significantly more diverse set of researchers in the coming years and decades. Bringing a broader range of perspective into the field will likely translate directly into more effective and fair artificial intelligence systems.

AN EXISTENTIAL THREAT FROM
SUPERINTELLIGENCE AND THE "CONTROL PROBLEM"

The AI risk that transcends all others is the possibility that machines with superhuman intelligence might someday wrest themselves from our direct control and pursue a course of action that ultimately presents an existential threat to humanity. Security issues, weaponization and algorithmic bias all pose immediate or near-term dangers. These are concerns that we clearly need to be addressing right now—before it is too late. An existential threat from superintelligence, however, is far more speculative and almost certainly lies decades—or perhaps even a century or more—in the future. Nonetheless, it is this risk that has captured the imagination of many prominent people and has received an enormous amount of media hype and attention.

The specter of existential AI risk emerged as a topic of serious public discussion in 2014. In May of that year, a group of scientists including the University of Cambridge cosmologist Stephen Hawking along with AI expert Stuart Russell and physicists Max Tegmark and Frank Wilczek co-authored an open letter published in the U.K.'s *Independent* declaring that the advent of artificial superintelligence "would be the biggest event in human history," and that a computer with superhuman intellectual capability might be capable of "outsmarting financial markets, out-inventing human researchers, out-manipulating human leaders, and developing weapons we cannot even understand." The letter warned that a failure to take this looming danger seriously might well turn out to be humanity's "worst mistake in history."[27]

Later that same year, the Oxford University philosopher Nick Bostrom published his book *Superintelligence: Paths, Dangers, Strategies*, which quickly became a somewhat surprising bestseller. Bostrom opens the book by pointing out that humans rule the earth purely on the basis of superior intellect. Many other animals are faster, stronger or more ferocious; it is our brains that led to dominance. Once another entity dramatically exceeds our own intellectual capability, the tables could easily be turned. As Bostrom puts it, "just as the fate of gorillas now depends more on us humans than on the gorillas themselves, so the fate of our species would depend on the actions of the machine superintelligence."[28]

Bostrom's book was enormously influential, especially among the Silicon Valley elite. Within a month of its publication, Elon Musk was declaring that "with artificial intelligence, we are summoning the demon" and that AI "could be more dangerous than nuclear weapons."[29] A year later, Musk would co-found OpenAI and give it the specific mission of building "friendly" artificial intelligence. Among those most deeply influenced by Bostrom's arguments, the idea that AI will someday pose an existential threat began to be perceived as a near certainly—and a danger ultimately far more terrifying and consequential than more mundane concerns like climate change or global pandemics. In a Ted Talk with more than five million views, the neuroscientist and philosopher Sam Harris argues that "it's very difficult to see how [the gains we make in artificial intelligence] won't destroy us or inspire us to destroy ourselves" and suggests that "we need something like a Manhattan Project" focused on avoiding that outcome by figuring out how to build friendly, controllable AI.[30]

None of this will be a concern, of course, until we manage to build a true thinking machine with cognitive capability at least equivalent to our own. As we saw in Chapter 5, the path to artificial general intelligence contains an unknown number of major hurdles, and it will likely take decades to achieve the necessary breakthroughs to reach this milestone. Recall that the mean estimate for AGI arrival from the top AI researchers I spoke to for my book *Architects of Intelligence* was about eighty years—or the end of this century. Once human-level AI becomes a reality, however, it is almost certain that superintelligence will rapidly follow. Indeed, any machine intelligence with the ability to learn and reason at the level of a human being would already be superior to us simply because it would also enjoy all the advantages that computers already have over us—including the ability to calculate and manipulate information at incomprehensible speed and to directly interface with other machines across networks.

Beyond this point, most AI experts assume that such a machine intelligence would soon decide to turn its intellectual energy toward improving its own design. This would then lead to relentless, recursive improvement as the system became ever smarter and more adept at re-engineering its own artificial mind. The result would inevitably be an "intelligence explosion"— a phenomenon that technoptimists like Ray Kurzweil believe will be the catalyst for the Singularity and the dawn of a new age. The argument that advances in AI would someday produce an explosion in machine intelligence was formulated long before Moore's Law began to deliver computer hardware that might bring such an event into the realm of possibility. In 1964, the mathematician I. J. Good wrote an academic paper entitled "Speculations Concerning the First Ultraintelligent Machine" in which he explained the concept like this:

> Let an ultraintelligent machine be defined as a machine that can far surpass all the intellectual activities of any man however clever. Since the design of machines is one of these intellectual activities, an ultraintelligent machine could design even better machines; there would then unquestionably be an "intelligence explosion," and the intelligence of man would be left far behind. Thus the first ultraintelligent machine is the *last* invention that man need ever make, provided that the machine is docile enough to tell us how to keep it under control.[31]

The promise that a superintelligent machine would be the last invention we ever need to make captures the optimism of Singularity proponents. The qualification that the machine must remain docile enough to be kept under control is the concern that suggests the possibility of an existential threat. This dark side of superintelligence is known in the AI community as the "control problem" or the "value alignment problem."

The control problem is not driven by fear of overtly malevolent machines of the kind portrayed in movies like *The Terminator*. Every AI system is designed around an objective function, in other words, a specific goal, expressed in mathematical terms, that the system strives to achieve. The concern is that a superintelligent system, given such an objective, might relentlessly pursue it using means that have unintended or unanticipated consequences that could turn out to be detrimental or even fatal to our civilization. A thought experiment involving a "paperclip maximizer" is often used to illustrate this point. Imagine a superintelligence designed with the specific objective of optimizing paperclip production. As it relentlessly pursued this goal, a superintelligent machine might invent new technologies that would allow it to convert virtually all the resources on earth into paperclips. Because the system would be so far beyond us in terms of its intellectual capability, it would likely be able to successfully foil any attempt to shut it down or alter its course of action. Indeed, any attempt at interference would be at odds with the system's objective function, and it would have a clear incentive to prevent this.

This example is obviously intended as a kind of cartoon. The real scenarios that might unfold in the future would likely be far more subtle, and the potential consequences would be much more difficult—or perhaps impossible—to anticipate in advance. We can already point to one important example that illustrates how unintended consequences can clearly be detrimental to the social fabric. The machine learning algorithms utilized by tech companies like YouTube and Facebook have generally been given the objective of maximizing user engagement on the platform. This in turn leads to more revenue from online advertisements. However, it has become evident that the algorithms pursuing this objective soon figured out that the best way to keep people engaged is to feed them ever more politically polarized content or tap directly into emotions like outrage or fear. This, for example, has led to the often-noted "rabbit hole" phenomenon on YouTube, in which a moderate video is followed by successive recommendations for ever more extreme content, all of which leads to sustained emotion-driven engagement with the platform.[32] That may be good for profitability, but it's clearly not good for our social or political environment. If a similar miscalculation were made with a superintelligent system, it might well be impossible to regain control as it sought to pursue its objective.

The quest to find a solution to the control problem has become an important topic of academic research at universities and especially within specialized, privately funded organizations such as OpenAI, Oxford University's Future of Humanity Institute, which is directed by Nick Bostrom, and the Machine Intelligence Research Institute located in Berkeley, California. In his 2019 book *Human Compatible: Artificial Intelligence and the Problem of Control*, Stuart Russell argues that the best solution to the problem is to not build an explicit objective function into advanced AI systems at all. Instead, systems should be designed to "maximize the realization of human preferences."[33] Because the machine intelligence could never be certain what these preferences or intentions are, it would have to formulate its objectives by studying human behavior and would be willing to dialogue with and accept guidance from humans. Unlike the unstoppable paperclip maximizer, such a system would submit to being shut down if it believed this was in line with the human preferences it was designed to optimize.

This represents a stark departure from the current approach to building AI systems. As Russell explains:

> Actually putting a model like this into practice requires a great deal of research. We need "minimally invasive" algorithms for decision making that prevent machines from messing with parts of the world whose value they are unsure about, as well as machines that learn more about our true, underlying preferences for how the future should unfold. Such machines will then face an age-old problem of moral philosophy: how to apportion benefits and costs among different individuals with conflicting desires.
>
> All this could take a decade to complete—and even then, regulations will be required to ensure provably safe systems are adopted while those that don't conform are retired. This won't be easy. But it's clear that this model must be in place *before* the abilities of A.I. systems exceed those of humans in the areas that matter.[34]

It's notable that with the exception of Stuart Russell, who is a co-author of the leading university artificial intelligence textbook, nearly all of the most prominent voices warning of a potential existential threat come from outside the fields of AI research or computer science. Instead, the alarm is primarily being sounded by public intellectuals like Sam Harris, Silicon Valley titans like Musk or scientists in other fields like Hawking or the MIT physicist Max Tegmark. Most of the experts engaged in actual AI research tend to be more sanguine. When I interviewed twenty-three elite researchers for my book *Architects of Intelligence*, I found that while a few took the possibility of an existential threat seriously, the vast majority were quite dismissive.

A common refrain is that the emergence of superintelligence is so far off, and the specific parameters of the problem to be solved so nebulous, that there's little point in pursuing the issue. Andrew Ng, who led AI research groups at Google and Baidu, is famous for saying that worrying about an existential threat from AI is like worrying about overpopulation on Mars—long before even the first team of astronauts has been sent to the red planet. The roboticist Rodney Brooks echoes this sentiment, saying that superintelligence is so far in the future that "it's not going to be a case of having exactly the same world as it is today, but with an AI superintelligence in the middle of it. . . . We have no clue at all about what the world or [a superintelligent AI system] are going to be like. Predicting an AI future is just a power game for isolated academics who live in a bubble away from the real world. That's not to say that these technologies aren't coming, but we won't know what they will look like before they arrive."[35]

Advocates for taking an existential AI threat seriously push back strongly against the idea that the issue is unimportant or unapproachable simply because it likely will not arise until decades have passed. They point out that the control problem needs to be solved *before* the first superintelligence comes into existence—or it will be too late. Stuart Russell likes to make an analogy to the arrival of extraterrestrials. Imagine we received a signal from space announcing that the aliens will be here in fifty years. Presumably we would immediately put in place a major global effort to prepare for the event. Russell believes we should be doing the same for the eventual arrival of superintelligence.

My own view is that the potential for an existential AI threat should be taken seriously. I think it's a very positive thing that researchers at organizations like the Future of Humanity Institute are actively working on the problem. However, it seems to me that this represents an appropriate allocation of resources, and that, for now at least, the issue is best addressed in a quiet academic research setting. It would be very difficult to justify anything on the scale of a government-funded "Manhattan Project" at this point in time. Nor does it seem wise to attempt to inject the issue into an already dysfunctional political process. Do we really want politicians with little or no understanding of the technology tweeting about the dangers of superintelligent machines? Given the very limited ability of the U.S. government in particular to accomplish almost anything at all, I also worry that hyping or politicizing a futuristic existential threat would be a distraction from the very real and immediate AI risks—including weaponization, security and bias—in which we really do need to begin investing significant resources in addressing right now.

A CRITICAL NEED FOR REGULATION

If there's one takeaway from the risks we've looked at in this chapter, it's that there is clearly an important role for government regulation as AI continues to advance and become more ubiquitous. However, I think it would be very misguided to overly regulate or place limits on general research into artificial intelligence. Doing so would likely be ineffective on a global basis because the research is taking place all over the world. And, as we've seen, China in particular is engaged in intense competition with the United States and other Western countries in pushing the AI frontier forward. Placing restraints on basic research would clearly put us at a significant disadvantage, and we simply cannot afford to fall behind China in the quest to be at the leading edge of such a consequential technology.

Rather, the focus should be on regulating specific applications of artificial intelligence. In areas such as self-driving cars or AI medical diagnostic tools, rules are already being developed because these applications intersect with a regulatory framework that is already in place. However, we need much broader oversight. Artificial intelligence will eventually touch virtually everything, and as we've seen, technologies like facial recognition or algorithms used in the criminal justice system are being used to make very high stakes decisions with virtually no guarantee that the technology is being deployed either effectively or justly.

Given the speed at which artificial intelligence is advancing and the complexity of the issues involved, I think it is unrealistic to expect the U.S. Congress, or indeed any parliamentary body, to write and enact detailed regulations in a timely fashion. The best course of action will probably be to create an independent governmental agency with regulatory powers specifically focused on applications of artificial intelligence. This would be an agency roughly comparable to the U.S. Food and Drug Administration, the Federal Aviation Administration or the Securities and Exchange Commission. In each case, these agencies—as well as their counterparts elsewhere, such as the European Medicines Agency—have developed deep in-house expertise that allows them to address the issues within their purview. The same needs to be true for the field of artificial intelligence. An AI regulatory agency would be given a broad mandate and allocated funds by Congress, but it would have the authority to write specific regulations and would be able to do this far more rapidly and effectively than the legislature.

Those with a libertarian orientation might well object and rightly point out that such an agency would suffer from the same inefficiencies already present in the rest of our regulatory apparatus. An AI regulatory agency would certainly have close relationships with large technology companies, we would likely see the proverbial "revolving-door," in which people move between industry and government, and there would be a significant risk of regulatory capture and undue influence on the part of the technology industry. These concerns are real, but nonetheless I think that such an agency is pretty clearly the optimal solution available to us. If the alternative is to simply do nothing, that surely will be far worse. In fact, a close relationship between the regulatory agency and the companies developing and deploying AI technology is likely to be as much a feature as a bug. Because government cannot realistically compete for top AI talent by offering the kind of salaries and equity compensation common in the technology industry, cooperation with the private sector may well be the only way the agency will be able to keep pace with the latest developments in the field. No solution will be perfect, but a productive alliance between industry, academia and government, centered in a regulatory agency with sufficient in-house expertise to keep things moving in the right direction, would go a long way toward ensuring that AI is deployed safely, inclusively and justly.

TWO AI FUTURES

AS ARTIFICIAL INTELLIGENCE CONTINUES TO ADVANCE AND EXTEND its reach into ever more facets of our lives, the risks associated with the technology will demand urgent attention. The intersection of the coronavirus crisis and widespread social upheaval in 2020 brought developments that suggest at least some of these issues are beginning to take a prominent place within public discourse. In the wake of the nationwide protests surrounding the killing of George Floyd by Minneapolis police officers in May, awareness of racial bias in facial recognition technology came to the forefront, and Amazon announced a one-year moratorium on sales of its Rekognition system to law enforcement agencies in order to give the U.S. Congress time to consider regulations on the technology. Microsoft announced a similar hiatus until legislation is passed, and IBM withdrew from the facial recognition market entirely.[1]

The coronavirus pandemic has also brought about a new openness to unconventional policy responses. As the shutdown of the economy led to staggering job losses, Congress was able to rapidly enact policies that would have been dead on arrival just a few months earlier. These included $1,200 stimulus payments sent directly to taxpayers, a dramatic, albeit temporary, increase in unemployment insurance payments and an expansion of the program to include gig economy workers. All these ideas will now be on the table as the impact of artificial intelligence and robotics on the job market accelerates in the coming years. Indeed, there have already been calls for monthly payments—essentially a basic income—to be paid for the duration of the crisis.[2]

Still, a far more comprehensive and cohesive response to the dangers that will inevitably arrive with the continued rise of AI is critical. This will require effective coordination between government and the private sector and the creation of a regulatory framework coupled with the expertise necessary to respond to rapid advances in the field. And all this needs to begin now, as we are arguably already behind the curve.

Despite these very real concerns, I firmly believe that the benefits from artificial intelligence will far outweigh the risks. Indeed, given the challenges we will face in the coming decades, I think AI will be indispensable. We will need artificial intelligence to launch us off our technological plateau into a new age of broad-based innovation.

Climate change looms as the most clearly foreseeable threat. In 2018, the Intergovernmental Panel on Climate Change released an analysis indicating that in order to keep global temperatures from increasing by more than 1.5 degrees Celsius—a threshold that will hopefully prevent catastrophic harm—we will need to cut net carbon emissions to zero by the year 2050. And in order to have any realistic chance of achieving this, we need roughly a forty-five percent reduction by 2030.[3]

The magnitude of this challenge was brought into stark relief by the massive and unprecedented experiment we conducted as the coronavirus pandemic emerged. As Bill Gates pointed out in an August 2020 blog post, the global shutdown—in which air travel came to a near halt and streets, highways and office buildings emptied across the globe—led to only about an eight percent reduction in emissions. And that temporary decrease came at a cost of untold trillions of dollars and skyrocketing unemployment in nearly every country on earth. In other words, assuming that we can somehow cut carbon emissions nearly in half over the next decade by relying primarily on policies that focus on conservation or behavioral changes such as shifting our commute to public transit seems unrealistic to say the least. As Gates says, "We cannot get to zero emissions simply—or even mostly—by flying and driving less."[4]

Success will depend first and foremost on innovation. And simply shifting to clean, renewable ways to generate electricity and power vehicles will not be sufficient. Electric power plants and transportation account for only about forty percent of global emissions. The remainder comes from agriculture, manufacturing, emissions from buildings and other miscellaneous sources.[5] Dramatically reducing global emissions will require technological breakthroughs in all these areas. Add in other challenges, such as the emerging global fresh water crisis or the inevitable next pandemic, and it becomes clear that we are desperately in need of a burst of innovation across the board. Yet, as we saw in Chapter 3, the pace of new idea creation has actually been slowing over the past few decades. As the economists at Stanford and MIT who studied innovation in the United States wrote, "Everywhere we look we find that ideas, and the exponential growth they imply, are getting harder to find."[6]

This has to change, and artificial intelligence is the catalyst that can make it happen. In the face of these challenges, nothing could be more consequential than a ubiquitous and affordable utility that dramatically amplifies human intellect and creativity. The key objective is to do everything possible to accelerate the development of this new resource while at the same time evolving our social safety net and regulatory framework in ways that will allow us to mitigate the accompanying risks and ensure that the dividends from AI are shared widely and inclusively.

As we navigate this path forward, I think that the future we build may ultimately fall somewhere on a spectrum bounded by two fictional extremes. The most optimistic scenario comes from the television show *Star Trek*. In this post-scarcity world, advanced technology has created material abundance, eliminated poverty, addressed environmental concerns and cured most disease. No one needs to toil in drudgery at an unrewarding job simply to sustain his or her survival. People in this world are highly educated and pursue challenges that they find rewarding. The absence of a traditional job has not led to idleness or a lack of meaning and dignity. In the *Star Trek* universe, people are valued for their intrinsic humanity—not primarily based on their economic output. Though many of the technologies portrayed in *Star Trek* are likely unrealizable or, at a minimum, lie in the distant future, I think the show offers a reasonable rendition of a future in which advanced technology leads to broad-based prosperity, solves humanity's terrestrial challenges and allows us to reach for the stars.

The alternate, and far more dystopian, future might be something closer to *The Matrix*. My fear is not that artificial intelligence will somehow enslave us, but rather that the real world might become so unequal, and so lacking in opportunity for most typical people to advance their prospects, that a large fraction of the population will choose to escape into alternative realities. As both AI and virtual reality technology accelerate over the coming years and decades, they will likely combine to create extraordinarily compelling and realistic simulations that, to many people, may seem far superior to the world in which they actually live. Indeed, a 2017 analysis by a group of economists found that the increasing number of young men who are detached from labor markets are spending an outsized portion of their time playing video games.[7] The technology will soon arrive to make these virtual environments so addictive that they might reasonably be viewed as a kind of drug.

If artificial intelligence and robotics upend the job market and employment opportunities evaporate or decline in quality, governments will in all likelihood eventually be forced to provide some form of support—perhaps a basic income—to citizens in order to maintain social order. If they neglect, however, to also ensure that the population continues to prioritize education and maintain a sense of purpose, the result is likely to be widespread apathy and disengagement. We might trend toward a society that balkanizes into a small elite that remains anchored in the real world, while the masses increasingly escape into a technological fantasy or perhaps are drawn to crime or other forms of addiction. We would then end up with a less educated population, a far less inclusive and effective democracy and a slower pace of innovation because many of the brightest individuals, lured into an ever more compelling virtual realm, might no longer see a strong incentive to strive for real-world success. In this scenario, economic and social headwinds will make it far more difficult for us to overcome the global challenges we face.

I think nearly everyone would agree that we should strive for a future that comes closer to *Star Trek*. This, however, will not happen by default. We will need to craft explicit policies designed to shift our trajectory toward that destination. In all likelihood, it will be a very long time before we arrive, but if we can begin by solving the problem of income distribution while maintaining a strong incentive for people to educate themselves and pursue meaningful challenges, we will be headed in the right direction.

ACKNOWLEDGMENTS

A great many people contributed to my understanding of artificial intelligence through conversations and technology demonstrations over the past few years. I'm especially grateful to the twenty-three prominent researchers and entrepreneurs who participated in the conversations recorded in *Architects of Intelligence*. They are truly among the brightest minds in the field of AI, and their insights and predictions informed much of the material in this book.

My editors, TJ Kelleher in the United States and Sarah Caro in the United Kingdom, were instrumental in helping me to refine my arguments and structure the manuscript into its optimal form. My agent, Don Fehr, once again found the proper home for this project at Basic Books.

The roughly eight months that I spent writing this book coincided with the emergence of the coronavirus pandemic and the ensuing shutdown. I was very fortunate to be able to remain safely at home and focus on my writing during this period, and I am profoundly grateful to all the health care professionals and frontline workers who had no such luxury.

Finally, I thank my wife, Xiaoxiao, and my daughter, Elaine, for their encouragement and support as I immersed myself in this project.

NOTES

CHAPTER 1. THE EMERGING DISRUPTION

1. Ewen Callaway, "'It will change everything': DeepMind's AI makes gigantic leap in solving protein structures," *Nature*, November 30, 2020, www.nature.com/articles/d41586-020-03348-4.

2. Andrew Senior, Demis Hassabis, John Jumper and Pushmeet Kohli, "AlphaFold: Using AI for scientific discovery," DeepMind Research Blog, January 15, 2020, deepmind.com/blog/article/AlphaFold-Using-AI-for -scientific-discovery.

3. Ian Sample, "Google's DeepMind predicts 3D shapes of proteins," *The Guardian*, December 2, 2018, www.theguardian.com/science/2018/dec/02 /google-deepminds-ai-program-alphafold-predicts-3d-shapes-of-proteins.

4. Lyxor Robotics and AI UCITS ETF, stock market ticker ROAI.

5. See, for example: Carl Benedikt Frey and Michael Osborne, "The future of employment: How susceptible are jobs to computerisation?," Oxford Martin School, University of Oxford, Working Paper, September 17, 2013, www .oxfordmartin.ox.ac.uk/downloads/academic/future-of-employment.pdf, p. 38.

6. Matt McFarland, "Elon Musk: 'With artificial intelligence we are summoning the demon,'" *Washington Post*, October 24, 2014, www.washington post.com/news/innovations/wp/2014/10/24/elon-musk-with-artificial-intelligence -we-are-summoning-the-demon/.

7. Anand S. Rao and Gerard Verweij, "Sizing the prize: What's the real value of AI for your business and how can you capitalise?," PwC, October 2018, www.pwc.com/gx/en/issues/analytics/assets/pwc-ai-analysis-sizing-the -prize-report.pdf.

CHAPTER 2. AI AS THE NEW ELECTRICITY

1. "Neuromorphic computing," Intel Corporation, accessed May 3, 2020, www.intel.com/content/www/us/en/research/neuromorphic-computing.html.

2. Sara Castellanos, "Intel to release neuromorphic-computing system," *Wall Street Journal*, March 18, 2020, www.wsj.com/articles/intel-to-release-neuromorphic-computing-system-11584540000.

3. Linda Hardesty, "WikiLeaks publishes the location of Amazon's data centers," SDXCentral, October 12, 2018, www.sdxcentral.com/articles/news/wikileaks-publishes-the-location-of-amazons-data-centers/2018/10/.

4. "RightScale 2019 State of the Cloud Report from Flexera," Flexera, 2019, resources.flexera.com/web/media/documents/rightscale-2019-state-of-the-cloud-report-from-flexera.pdf, p. 2.

5. Pierr Johnson, "With the public clouds of Amazon, Microsoft and Google, big data is the proverbial big deal," *Forbes*, June 15, 2017, www.forbes.com/sites/johnsonpierr/2017/06/15/with-the-public-clouds-of-amazon-microsoft-and-google-big-data-is-the-proverbial-big-deal/.

6. Richard Evans and Jim Gao, "DeepMind AI reduces Google data centre cooling bill by 40%," DeepMind Research Blog, July 20, 2016, deepmind.com/blog/article/deepmind-ai-reduces-google-data-centre-cooling-bill-40.

7. Urs Hölzle, "Data centers are more energy efficient than ever," Google Blog, February 27, 2020, www.blog.google/outreach-initiatives/sustainability/data-centers-energy-efficient/.

8. Ron Miller, "AWS revenue growth slips a bit, but remains Amazon's golden goose," *TechCrunch*, July 25, 2019, techcrunch.com/2019/07/25/aws-revenue-growth-slips-a-bit-but-remains-amazons-golden-goose/.

9. John Bonazzo, "Google exits Pentagon 'JEDI' project after employee protests," *Observer*, October 10, 2018, observer.com/2018/10/google-pentagon-jedi/.

10. Annie Palmer, "Judge temporarily blocks Microsoft Pentagon cloud contract after Amazon suit," CNBC, February 13, 2020, www.cnbc.com/2020/02/13/amazon-gets-restraining-order-to-block-microsoft-work-on-pentagon-jedi.html.

11. Lauren Feiner, "DoD asks judge to let it reconsider decision to give Microsoft $10 billion contract over Amazon," CNBC, March 13, 2020, www.cnbc.com/2020/03/13/pentagon-asks-judge-to-let-it-reconsider-its-jedi-cloud-contract-award.html.

12. "TensorFlow on AWS," Amazon Web Services, accessed May 4, 2020, aws.amazon.com/tensorflow/.

13. Kyle Wiggers, "Intel debuts Pohoiki Springs, a powerful neuromorphic research system for AI workloads," *VentureBeat*, March 18, 2020, venturebeat.com/2020/03/18/intel-debuts-pohoiki-springs-a-powerful-neuromorphic-research-system-for-ai-workloads/.

14. Jeremy Kahn, "Inside big tech's quest for human-level A.I.," *Fortune*, January 20, 2020, fortune.com/longform/ai-artificial-intelligence-big-tech-microsoft-alphabet-openai/.

15. Martin Ford, Interview with Fei-Fei Li, in *Architects of Intelligence: The Truth about AI from the People Building It*, Packt Publishing, 2018, p. 150.

16. "Deep Learning on AWS," Amazon Web Services, accessed May 4, 2020, aws.amazon.com/deep-learning/.

17. Kyle Wiggers, "MIT researchers: Amazon's Rekognition shows gender and ethnic bias," *VentureBeat*, January 24, 2019, venturebeat.com/2019/01 /24/amazon-rekognition-bias-mit/.

18. "New schemes teach the masses to build AI," *The Economist*, October 27, 2018, www.economist.com/business/2018/10/27/new-schemes -teach-the-masses-to-build-ai.

19. Chris Hoffman, "What is 5G, and how fast will it be?," *How-to Geek*, January 3, 2020, www.howtogeek.com/340002/what-is-5g-and-how -fast-will-it-be/.

CHAPTER 3. BEYOND HYPE

1. Tesla, "Tesla Autonomy Day (video)," YouTube, April 22, 2019, www.youtube.com/watch?reload=9&v=Ucp0TTmvqOE.

2. Sean Szymkowski, "Tesla's full self-driving mode under the watchful eye of NHTSA," *Road Show*, October 22, 2020, www.cnet.com/roadshow/news /teslas-full-self-driving-mode-nhtsa/.

3. Rob Csongor, "Tesla raises the bar for self-driving carmakers," NVIDIA Blog, April 23, 2019, blogs.nvidia.com/blog/2019/04/23/tesla-self-driving/.

4. Jeffrey Van Camp, "My Jibo is dying and it's breaking my heart," *Wired*, March 9, 2019, www.wired.com/story/jibo-is-dying-eulogy/.

5. Mark Gurman and Brad Stone, "Amazon is said to be working on another big bet: Home robots," *Bloomberg*, April 23, 2018, www .bloomberg.com/news/articles/2018-04-23/amazon-is-said-to-be-working -on-another-big-bet-home-robots.

6. Martin Ford, Interview with Rodney Brooks, in *Architects of Intelligence: The Truth about AI from the People Building It*, Packt Publishing, 2018, p. 432.

7. "Solving Rubik's Cube with a robot hand," OpenAI, October 15, 2019, openai.com/blog/solving-rubiks-cube/. (Includes videos.)

8. Will Knight, "Why solving a Rubik's Cube does not signal robot supremacy," *Wired*, October 16, 2019, www.wired.com/story/why-solving -rubiks-cube-not-signal-robot-supremacy/.

9. Noam Scheiber, "Inside an Amazon warehouse, robots' ways rub off on humans," *New York Times*, July 3, 2019, www.nytimes.com/2019/07/03 /business/economy/amazon-warehouse-labor-robots.html.

10. Eugene Kim, "Amazon's $775 million deal for robotics company Kiva is starting to look really smart," *Business Insider*, June 15, 2016, www .businessinsider.com/kiva-robots-save-money-for-amazon-2016-6.

11. Will Evans, "Ruthless quotas at Amazon are maiming employees," *The Atlantic*, November 25, 2019, www.theatlantic.com/technology /archive/2019/11/amazon-warehouse-reports-show-worker-injuries/602530/.

12. Jason Del Ray, "How robots are transforming Amazon warehouse jobs—for better and worse," *Recode*, December 11, 2019, www.vox.com /recode/2019/12/11/20982652/robots-amazon-warehouse-jobs-automation.

13. Michael Sainato, "'I'm not a robot': Amazon workers condemn unsafe, grueling conditions at warehouse," *The Guardian*, February 5, 2020, www.theguardian.com/technology/2020/feb/05/amazon-workers-protest -unsafe-grueling-conditions-warehouse.

14. Jeffrey Dastin, "Exclusive: Amazon rolls out machines that pack orders and replace jobs," Reuters, May 13, 2019, www.reuters.com/article/us -amazon-com-automation-exclusive/exclusive-amazon-rolls-out-machines -that-pack-orders-and-replace-jobs-idUSKCN1SJ0X1.

15. Matt Simon, "Inside the Amazon warehouse where humans and machines become one," *Wired*, June 5, 2019, www.wired.com/story/amazon -warehouse-robots/.

16. James Vincent, "Amazon's latest robot champion uses deep learning to stock shelves," *The Verge*, July 5, 2016, www.theverge.com/2016 /7/5/12095788/amazon-picking-robot-challenge-2016.

17. Jeffrey Dastin, "Amazon's Bezos says robotic hands will be ready for commercial use in next 10 years," Reuters, June 6, 2019, www.reuters.com /article/us-amazon-com-conference/amazons-bezos-says-robotic-hands -will-be-ready-for-commercial-use-in-next-10-years-idUSKCN1T72JB.

18. Tech Insider, "Inside a warehouse where thousands of robots pack groceries (video)," YouTube, May 9, 2018, www.youtube.com/watch ?reload=9&v=4DKrcpa8Z_E.

19. James Vincent, "Welcome to the automated warehouse of the future," *The Verge*, May 8, 2018, www.theverge.com/2018/5/8/17331250 /automated-warehouses-jobs-ocado-andover-amazon.

20. Ibid.

21. "ABB and Covariant partner to deploy integrated AI robotic solutions," ABB Press Release, February 25, 2020, new.abb.com/news/detail/57457 /abb-and-covariant-partner-to-deploy-integrated-ai-robotic-solutions.

22. Evan Ackerman, "Covariant uses simple robot and gigantic neural net to automate warehouse picking," *IEEE Spectrum*, January 29, 2020, spec trum.ieee.org/automaton/robotics/industrial-robots/covariant-ai-gigantic -neural-network-to-automate-warehouse-picking.

23. Jonathan Vanian, "Industrial robotics giant teams up with a rising A.I. startup," *Fortune*, February 25, 2020, fortune.com/2020/02/25 /industrial-robotics-ai-covariant/.

24. Alexander Lavin, J. Swaroop Guntupalli, Miguel Lázaro-Gredilla, et al., "Explaining visual cortex phenomena using recursive cortical network," Vicarious Research Paper, July 30, 2018, www.biorxiv.org/content/biorxiv /early/2018/07/30/380048.full.pdf.

25. Tom Simonite, "These industrial robots get more adept with every task," *Wired*, March 10, 2020, www.wired.com/story/these-industrial -robots-adept-every-task/.

26. Adam Satariano and Cade Metz, "A warehouse robot learns to sort out the tricky stuff," *New York Times*, January 29, 2020, www.nytimes.com/2020/01/29/technology/warehouse-robot.html.

27. Matthew Boyle, "Robots in aisle two: Supermarket survival means matching Amazon," *Bloomberg*, December 3, 2019, www.bloomberg.com/features/2019-automated-grocery-stores/.

28. Ibid.

29. Nathaniel Meyersohn, "Grocery stores turn to robots during the coronavirus," CNN Business, April 7, 2020, www.cnn.com/2020/04/07/business/grocery-stores-robots-automation/index.html.

30. Shoshy Ciment, "Walmart is bringing robots to 650 more stores as the retailer ramps up automation in stores nationwide," *Business Insider*, January 13, 2020, www.businessinsider.com/walmart-adding-robots-help-stock-shelves-to-650-more-stores-2020-1.

31. Jennifer Smith, "Grocery delivery goes small with micro-fulfillment centers," *Wall Street Journal*, January 27, 2020, www.wsj.com/articles/grocery-delivery-goes-small-with-micro-fulfillment-centers-11580121002.

32. Nick Wingfield, "Inside Amazon Go, a store of the future," *New York Times*, January 21, 2018, www.nytimes.com/2018/01/21/technology/inside-amazon-go-a-store-of-the-future.html.

33. Spencer Soper, "Amazon will consider opening up to 3,000 cashierless stores by 2021," *Bloomberg*, September 29, 2018, www.bloomberg.com/news/articles/2018-09-19/amazon-is-said-to-plan-up-to-3-000-cashierless-stores-by-2021.

34. Paul Sawyers, "SoftBank leads $30 million investment in Accel Robotics for AI-enabled cashierless stores," *VentureBeat*, December 3, 2019, venturebeat.com/2019/12/03/softbank-leads-30-million-investment-in-accel-robotics-for-ai-enabled-cashierless-stores/.

35. Jurica Dujmovic, "As coronavirus hits hard, Amazon starts licensing cashier-free technology to retailers," MarketWatch, March 31, 2020, www.marketwatch.com/story/as-coronavirus-hits-hard-amazon-starts-licensing-cashier-free-technology-to-retailers-2020-03-31.

36. Eric Rosenbaum, "Panera is losing nearly 100% of its workers every year as fast-food turnover crisis worsens," CNBC, August 29, 2019, www.cnbc.com/2019/08/29/fast-food-restaurants-in-america-are-losing-100percent-of-workers-every-year.html.

37. Ibid.

38. Kate Krader, "The world's first robot-made burger is about to hit the Bay Area," *Bloomberg*, June 21, 2018, www.bloomberg.com/news/features/2018-06-21/the-world-s-first-robotic-burger-is-ready-to-hit-the-bay-area.

39. John Elflein, "U.S. health care expenditure as a percentage of GDP 1960–2020," Statista, June 8, 2020, www.statista.com/statistics/184968/us-health-expenditure-as-percent-of-gdp-since-1960/.

40. "Healthcare expenditure and financing," OCED.stat, accessed May 15, 2020, stats.oecd.org/Index.aspx?DataSetCode=SHA.

41. William J. Baumol and William G. Bowen, *Performing Arts, The Economic Dilemma: A Study of Problems Common to Theater, Opera, Music and Dance*, MIT Press, 1966.

42. Michael Maiello, "Diagnosing William Baumol's cost disease," *Chicago Booth Review*, May 18, 2017, review.chicagobooth.edu/economics/2017/article/diagnosing-william-baumol-s-cost-disease.

43. "7 healthcare robots for the smart hospital of the future," Nanalyze, April 6, 2020, www.nanalyze.com/2020/04/healthcare-robots-smart-hospital/.

44. Daphne Sashin, "Robots join workforce at the new Stanford Hospital," *Stanford Medicine News*, November 4, 2019, med.stanford.edu/news/all-news/2019/11/robots-join-the-workforce-at-the-new-stanford-hospital-.html.

45. Diego Ardila, Atilla P. Kiraly, Sujeeth Bharadwaj et al., "End-to-end lung cancer screening with three-dimensional deep learning on low-dose chest computed tomography," *Nature Medicine*, volume 25, pp. 954–961 (2019), May 20, 2019, www.nature.com/articles/s41591-019-0447-x.

46. Karen Hao, "Doctors are using AI to triage COVID-19 patients. The tools may be here to stay," *MIT Technology Review*, April 23, 2020, www.technologyreview.com/2020/04/23/1000410/ai-triage-covid-19-patients-health-care.

47. Creative Distribution Lab, "Geoffrey Hinton: On radiology (video)," YouTube, November 24, 2016, www.youtube.com/watch?reload=9&v=2HM PRXstSvQ. (Part of the Machine Learning and the Market for Intelligence 2016 conference.)

48. Alex Bratt, "Why radiologists have nothing to fear from deep learning," *Journal of the American College of Radiology*, volume 16, issue 9, Part A, pp. 1190–1192 (September 2019), April 18, 2019, www.jacr.org/article/S1546-1440(19)30198-X/fulltext.

49. Ray Sipherd, "The third-leading cause of death in US most doctors don't want you to know about," CNBC, February 22, 2018, www.cnbc.com/2018/02/22/medical-errors-third-leading-cause-of-death-in-america.html.

50. Elise Reuter, "Study shows reduction in medication errors using health IT startup's software," *MedCity News*, December 24, 2019, medcitynews.com/2019/12/study-shows-reduction-in-medication-errors-using-health-it-startups-software/.

51. Adam Vaughan, "Google is taking over DeepMind's NHS contracts—should we be worried?," *New Scientist*, September 27, 2019, www.newscientist.com/article/2217939-google-is-taking-over-deepminds-nhs-contracts-should-we-be-worried/.

52. Clive Thompson, "May A.I. help you?," *New York Times*, November 14, 2018, www.nytimes.com/interactive/2018/11/14/magazine/tech-design-ai-chatbot.html.

53. Blair Hanley Frank, "Woebot raises $8 million for its AI therapist," *VentureBeat*, March 1, 2018, venturebeat.com/2018/03/01/woebot-raises-8 -million-for-its-ai-therapist/.

54. Ariana Eunjung Cha, "Watson's next feat? Taking on cancer," *Washington Post*, June 27, 2015, www.washingtonpost.com/sf/national/2015/06/27 /watsons-next-feat-taking-on-cancer/.

55. Mary Chris Jaklevic, "MD Anderson Cancer Center's IBM Watson project fails, and so did the journalism related to it," *Health News Review*, February 23, 2017, www.healthnewsreview.org/2017/02/md-anderson -cancer-centers-ibm-watson-project-fails-journalism-related/.

56. Mark Anderson, "Surprise! 2020 is not the year for self-driving cars," *IEEE Spectrum*, April 22, 2020, spectrum.ieee.org/transportation/self-driving /surprise-2020-is-not-the-year-for-selfdriving-cars.

57. Alex Knapp, "Aurora CEO Chris Urmson says there'll be hundreds of self-driving cars on the road in five years," *Forbes*, October 29, 2019, www.forbes.com/sites/alexknapp/2019/10/29/aurora-ceo-chris-urmson-says -therell-be-hundreds-of-self-driving-cars-on-the-road-in-five-years/.

58. Lex Fridman, "Chris Urmson: Self-driving cars at Aurora, Google, CMU, and DARPA," Artificial Intelligence Podcast, episode 28, July 22, 2019, lexfridman.com/chris-urmson/. (Video and audio podcast available.)

59. Stefan Seltz-Axmacher, "The end of Starsky Robotics," Starsky Robotics 10-4 Labs Blog, March 19, 2020, medium.com/starsky-robotics-blog /the-end-of-starsky-robotics-acb8a6a8a5f5.

60. Sam Dean, "Uber fares are cheap, thanks to venture capital. But is that free ride ending?," *Los Angeles Times*, May 11, 2019, www.latimes.com /business/technology/la-fi-tn-uber-ipo-lyft-fare-increase-20190511-story.html.

61. Darrell Etherington, "Waymo has now driven 10 billion autonomous miles in simulation," TechCrunch, July 10, 2019, techcrunch.com/2019/07/10 /waymo-has-now-driven-10-billion-autonomous-miles-in-simulation/.

62. Waymo website, accessed May 20, 2020, waymo.com/.

63. Ray Kurzweil, "The Law of Accelerating Returns," Kurzweil Library Blog, March 7, 2001, www.kurzweilai.net/the-law-of-accelerating-returns.

64. Tyler Cowen, *The Great Stagnation: How America Ate All the Low-Hanging Fruit of Modern History, Got Sick, and Will (Eventually) Feel Better*, Dutton, 2011.

65. Robert J. Gordon, *The Rise and Fall of American Growth: The U.S. Standard of Living Since the Civil War*, Princeton University Press, 2016.

66. Nicholas Bloom, Charles I. Jones, John Van Reenen and Michael Webb, "Are ideas getting harder to find?" *American Economic Review*, volume 110, issue 4, pp. 1104–1144 (April 2020), www.aeaweb.org/articles?id=10.1257 /aer.20180338, p. 1138.

67. Ibid., p. 1104.

68. Ibid., p. 1104.

69. Sam Lemonick, "Exploring chemical space: Can AI take us where no human has gone before?," *Chemical and Engineering News*, April 6, 2020, cen.acs.org/physical-chemistry/computational-chemistry/Exploring-chemical-space-AI-take/98/i13.

70. Ibid.

71. Delft University of Technology, "Researchers design new material using artificial intelligence," Phys.org, October 14, 2019, phys.org/news/2019-10-material-artificial-intelligence.html.

72. Beatrice Jin, "How AI helps to advance new materials discovery," Cornell Research, accessed May 22, 2020, research.cornell.edu/research/how-ai-helps-advance-new-materials-discovery.

73. Savanna Hoover, "Artificial intelligence meets materials science," Texas A&M University Engineering News, December 17, 2018, engineering.tamu.edu/news/2018/12/artificial-intelligence-meets-materials-science.html.

74. Kyle Wiggers, "Kebotix raises $11.5 million to automate lab experiments with AI and robotics," *VentureBeat*, April 16, 2020, venturebeat.com/2020/04/16/kebotix-raises-11-5-million-to-automate-lab-experiments-with-ai-and-robotics/.

75. Simon Smith, "230 startups using artificial intelligence in drug discovery," BenchSci Blog, updated April 8, 2020, blog.benchsci.com/startups-using-artificial-intelligence-in-drug-discovery.

76. Ford, Interview with Daphne Koller, in *Architects of Intelligence*, p. 388.

77. Ned Pagliarulo, "AI's impact in drug discovery is coming fast, predicts GSK's Hal Barron," *BioPharma Dive*, November 21, 2019, www.biopharmadive.com/news/gsk-hal-barron-ai-drug-discovery-prediction-daphne-koller/567855/.

78. Anne Trafton, "Artificial intelligence yields new antibiotic," *MIT News*, February 20, 2020, news.mit.edu/2020/artificial-intelligence-identifies-new-antibiotic-0220.

79. Richard Staines, "Exscientia claims world first as AI-created drug enters clinic," *Pharmaphorum*, January 30, 2020, pharmaphorum.com/news/exscientia-claims-world-first-as-ai-created-drug-enters-clinic/.

80. Matt Reynolds, "DeepMind's AI is getting closer to its first big real-world application," *Wired*, January 15, 2020, www.wired.co.uk/article/deepmind-protein-folding-alphafold.

81. Semantic Scholar website, accessed May 25, 2020, pages.semanticscholar.org/about-us.

82. Ibid.

83. Khari Johnson, "Microsoft, White House, and Allen Institute release coronavirus data set for medical and NLP researchers," *VentureBeat*, March 16, 2020, venturebeat.com/2020/03/16/microsoft-white-house-and-allen-institute-release-coronavirus-data-set-for-medical-and-nlp-researchers/.

84. "CORD-19: COVID-19 Open Research Dataset," Semantic Scholar, accessed May 6, 2020, www.semanticscholar.org/cord19.

CHAPTER 4. THE QUEST TO BUILD INTELLIGENT MACHINES

1. Samuel Butler, "Darwin among the machines, a letter to the editors," *The Press*, Christchurch, New Zealand, June 13, 1863.

2. Alan Turing, "Computing machinery and intelligence," *Mind*, volume LIX, issue 236, pp. 433–460 (October 1950).

3. J. McCarthy, M. L. Minsky, N. Rochester and C. E. Shannon, "A proposal for the Dartmouth Summer Research Project on Artificial Intelligence," August 31, 1955, raysolomonoff.com/dartmouth/boxa/dart564props.pdf.

4. Brad Darrach, "Meet Shaky, the first electronic person: The fascinating and fearsome reality of a machine with a mind of its own," *LIFE*, November 20, 1970, p. 58D.

5. Ibid.

6. Warren McCulloch and Walter Pitts, "A logical calculus of ideas immanent in nervous activity," *Bulletin of Mathematical Biophysics*, volume 5, issue 4, pp. 115–133 (December 1943).

7. Martin Ford, Interview with Ray Kurzweil, in *Architects of Intelligence: The Truth about AI from the People Building It*, Packt Publishing, 2018, p. 228.

8. Marvin Minsky and Seymour Papert, *Perceptrons: An Introduction to Computational Geometry*, MIT Press, 1969.

9. Ford, Interview with Yann LeCun, in *Architects of Intelligence*, p. 122.

10. David E. Rumelhart, Geoffrey E. Hinton and Ronald J. Williams, "Learning representations by back-propagating errors," *Nature*, volume 323, issue 6088, pp. 533–536 (1986), October 9, 1986, www.nature.com /articles/323533a0.

11. Ford, Interview with Geoffrey Hinton, in *Architects of Intelligence*, p. 73.

12. Dave Gershgorn, "The data that transformed AI research—and possibly the world," *Quartz*, July 26, 2017, qz.com/1034972/the-data-that -changed-the-direction-of-ai-research-and-possibly-the-world/.

13. Ford, Interview with Geoffrey Hinton, in *Architects of Intelligence*, p. 77.

14. Email from Jürgen Schmidhuber to Martin Ford, January 28, 2019.

15. Jürgen Schmidhuber, "Critique of paper by 'Deep Learning Conspiracy' (Nature 521 p 436)," June 2015, people.idsia.ch/~juergen/deep-learning -conspiracy.html.

16. John Markoff, "When A.I. matures, it may call Jürgen Schmidhuber 'Dad,'" *New York Times*, November 27, 2016, www.nytimes.com/2016/11/27 /technology/artificial-intelligence-pioneer-jurgen-schmidhuber-overlooked.html.

17. Robert Triggs, "What being an 'AI first' company means for Google," *Android Authority*, November 8, 2017, www.androidauthority.com/google -ai-first-812335/.

18. Cade Metz, "Why A.I. researchers at Google got desks next to the boss," *New York Times*, February 19, 2018, www.nytimes.com/2018/02/19 /technology/ai-researchers-desks-boss.html.

CHAPTER 5. DEEP LEARNING AND
THE FUTURE OF ARTIFICIAL INTELLIGENCE

1. Martin Ford, Interview with Geoffrey Hinton, in *Architects of Intelligence: The Truth about AI from the People Building It*, Packt Publishing, 2018, pp. 72–73.

2. Matt Reynolds, "New computer vision challenge wants to teach robots to see in 3D," *New Scientist*, April 7, 2017, www.newscientist.com /article/2127131-new-computer-vision-challenge-wants-to-teach-robots-to -see-in-3d/.

3. Ashlee Vance, "Silicon Valley's latest unicorn is run by a 22-year-old," *Bloomberg Businessweek*, August 5, 2019, www.bloomberg.com/news /articles/2019-08-05/scale-ai-is-silicon-valley-s-latest-unicorn.

4. Volodymyr Mnih, Koray Kavukcuoglu, David Silver et al. "Playing Atari with deep reinforcement learning," DeepMind Research, January 1, 2013, deepmind.com/research/publications/playing-atari-deep-reinforcement-learning.

5. Volodymyr Mnih, Koray Kavukcuoglu, David Silver et al., "Human-level control through deep reinforcement learning," *Nature*, volume 518, pp. 529– 533 (2015), February 25, 2015, www.nature.com/articles/nature14236.

6. Tu Yuanyuan, "The game of Go: Ancient wisdom," *Confucius Institute Magazine*, volume 17, pp. 46–51 (November 2011), confuciusmag.com /go-game.

7. David Silver and Demis Hassabis, "AlphaGo: Mastering the ancient game of Go with machine learning," Google AI Blog, January 27, 2016, ai.googleblog.com/2016/01/alphago-mastering-ancient-game-of-go.html.

8. Matt Schiavenza, "China's 'Sputnik Moment' and the Sino-American battle for AI supremacy," Asia Society Blog, September 25, 2018, asiasociety .org/blog/asia/chinas-sputnik-moment-and-sino-american-battle-ai-supremacy.

9. John Markoff, "Scientists see promise in deep-learning programs," *New York Times*, November 23, 2012, www.nytimes.com/2012/11/24/science /scientists-see-advances-in-deep-learning-a-part-of-artificial-intelligence.html.

10. Dario Amodei and Danny Hernandez, "AI and Compute," OpenAI Blog, May 16, 2018, openai.com/blog/ai-and-compute/.

11. Will Knight, "Facebook's head of AI says the field will soon 'hit the wall,'" *Wired*, December 4, 2019, www.wired.com/story/facebooks-ai-says -field-hit-wall/.

12. Kim Martineau, "Shrinking deep learning's carbon footprint," *MIT News*, August 7, 2020, news.mit.edu/2020/shrinking-deep-learning-carbon -footprint-0807.

13. "General game playing with schema networks," Vicarious Research, August 7, 2017, www.vicarious.com/2017/08/07/general-game-playing-with -schema-networks/.

14. Sam Shead, "Researchers: Are we on the cusp of an 'AI winter'?," BBC News, January 12, 2020, www.bbc.com/news/technology-51064369.

15. Filip Piekniewski, "AI winter is well on its way," Piekniewski's Blog, May 28, 2018, blog.piekniewski.info/2018/05/28/ai-winter-is-well-on-its-way/.

16. Ford, Interview with Jeffery Dean, in *Architects of Intelligence*, p. 377.

17. Ford, Interview with Demis Hassabis, in *Architects of Intelligence*, p. 171.

18. Andrea Banino, Caswell Barry, Dharshan Kumaran and Benigno Uria, "Navigating with grid-like representations in artificial agents," DeepMind Research Blog, May 9, 2018, deepmind.com/blog/article/grid-cells.

19. Ford, Interview with Demis Hassabis, in *Architects of Intelligence*, p. 173.

20. Andrea Banino, Caswell Barry, Benigno Uria et al., "Vector-based navigation using grid-like representations in artificial agents," *Nature*, volume 557, pp. 429–433 (2018), May 9, 2018, www.nature.com/articles /s41586-018-0102-6.

21. Will Dabney and Zeb Kurth-Nelson, "Dopamine and temporal difference learning: A fruitful relationship between neuroscience and AI," DeepMind Research Blog, January 15, 2020, deepmind.com/blog/article /Dopamine-and-temporal-difference-learning-A-fruitful-relationship-between -neuroscience-and-AI.

22. Tony Peng, "Yann LeCun Cake Analogy 2.0," *Synced Review*, February 22, 2019, medium.com/syncedreview/yann-lecun-cake-analogy-2-0-a361 da560dae.

23. Ford, Interview with Demis Hassabis, in *Architects of Intelligence*, pp. 172–173.

24. Jeremy Kahn, "A.I. breakthroughs in natural-language processing are big for business," *Fortune*, January 20, 2020, fortune.com/2020/01/20 /natural-language-processing-business/.

25. Ford, Interview with David Ferrucci, in *Architects of Intelligence*, p. 409.

26. Ibid. p. 414.

27. *Do You Trust This Computer?*, released April 5, 2018, Papercut Films, doyoutrustthiscomputer.org/.

28. Ford, Interview with David Ferrucci, in *Architects of Intelligence*, p. 414.

29. Ray Kurzweil, *The Singularity Is Near: When Humans Transcend Biology*, Penguin Books, 2005.

30. Ray Kurzweil, *How to Create a Mind: The Secret of Human Thought Revealed*, Penguin Books, 2012.

31. Ford, Interview with Ray Kurzweil, in *Architects of Intelligence*, pp. 230–231.

32. Mitch Kapor and Ray Kurzweil, "A wager on the Turing test: The rules," Kurzweil AI Blog, April 9, 2002, www.kurzweilai.net/a-wager-on-the-turing -test-the-rules.

33. Sean Levinson, "A Google executive is taking 100 pills a day so he can live forever," *Elite Daily*, April 15, 2015, www.elitedaily.com/news/world /google-executive-taking-pills-live-forever/1001270.

34. Ford, Interview with Ray Kurzweil, in *Architects of Intelligence*, pp. 240–241.

35. Ibid., p. 230.

36. Ibid., p. 233.

37. Alec Radford, Jeffrey Wu, Dario Amodei et al., "Better language models and their implications," OpenAI Blog, February 14, 2019, openai.com/blog /better-language-models/.

38. James Vincent, "OpenAI's latest breakthrough is astonishingly powerful, but still fighting its flaws," *The Verge*, July 30, 2020, www.theverge .com/21346343/gpt-3-explainer-openai-examples-errors-agi-potential.

39. Gary Marcus and Ernest Davis, "GPT-3, Bloviator: OpenAI's language generator has no idea what it's talking about," *MIT Technology Review*, August 22, 2020, www.technologyreview.com/2020/08/22/1007539 /gpt3-openai-language-generator-artificial-intelligence-ai-opinion/.

40. Ford, Interview with Stuart Russell, in *Architects of Intelligence*, p. 53.

41. "OpenAI Founder: Short-Term AGI Is a Serious Possibility," *Synced*, November 13, 2018, syncedreview.com/2018/11/13/openai-founder-short -term-agi-is-a-serious-possibility/.

42. Connie Loizos, "Sam Altman in conversation with StrictlyVC (video)," YouTube, May 18, 2019, youtu.be/TzcJlKg2Rc0, location 39:00.

43. Luke Dormehl, "Neuro-symbolic A.I. is the future of artificial intelligence. Here's how it works," *Digital Trends*, January 5, 2020, www .digitaltrends.com/cool-tech/neuro-symbolic-ai-the-future/.

44. Ford, Interview with Yoshua Bengio, in *Architects of Intelligence*, p. 22.

45. Ford, Interview with Geoffrey Hinton, in *Architects of Intelligence*, pp. 84–85.

46. Ford, Interview with Yann LeCun, in *Architects of Intelligence*, p. 123.

47. Anthony M. Zador, "A critique of pure learning and what artificial neural networks can learn from animal brains," *Nature Communications*, volume 10, article number 3770 (2019), August 21, 2019, www.nature.com/articles /s41467-019-11786-6.

48. Zoey Chong, "AI beats humans in Stanford reading comprehension test," CNET, January 16, 2018, www.cnet.com/news/new-results-show-ai-is -as-good-as-reading-comprehension-as-we-are/.

49. All Winograd schema examples are taken from: Ernest Davis, "A collection of Winograd schemas," New York University Department of Computer Science, September 8, 2011, cs.nyu.edu/davise/papers/WSOld.html.

50. Ford, Interview with Oren Etzioni, in *Architects of Intelligence*, pp. 495–496.

51. Ibid.

52. Ford, Interview with Yoshua Bengio, in *Architects of Intelligence*, p. 21.

53. Ford, Interview with Yann LeCun, in *Architects of Intelligence*, pp. 126–127.

54. Ibid., p. 130.

55. Ford, Interview with Judea Pearl, in *Architects of Intelligence*, p. 364.

56. Ford, Interview with Joshua Tenenbaum, in *Architects of Intelligence*, pp. 471–472.

57. Ford, Interview with Judea Pearl, in *Architects of Intelligence*, p. 366.

58. Will Knight, "An AI pioneer wants his algorithms to understand the 'why,'" *Wired*, October 8, 2019, www.wired.com/story/ai-pioneer-algorithms-understand-why/.

59. Graham Allison, *Destined for War: Can America and China Escape Thucydides's Trap?*, Houghton Mifflin Harcourt, 2017.

60. The AlphaStar team, "AlphaStar: Mastering the real-time strategy game *StarCraft II*," DeepMind Research Blog, January 24, 2019, deepmind.com /blog/article/alphastar-mastering-real-time-strategy-game-starcraft-ii.

61. Ford, Interview with Oren Etzioni, in *Architects of Intelligence*, p. 494.

62. Ford, *Architects of Intelligence*, p. 528.

63. "AI timeline surveys," AI Impacts, accessed June 27, 2020, aiimpacts.org /ai-timeline-surveys/.

CHAPTER 6. DISAPPEARING JOBS
AND THE ECONOMIC CONSEQUENCES OF AI

1. David Axelrod, "Larry Summers," The Axe Files (podcast), episode 98, November 21, 2016, omny.fm/shows/the-axe-files-with-david-axelrod/ep-98 -larry-summers.

2. Sam Fleming and Brooke Fox, "US states that voted for Trump most vulnerable to job automation," *Financial Times*, January 23, 2019, www .ft.com/content/cbf2a01e-1f41-11e9-b126-46fc3ad87c65.

3. Carol Graham, "Understanding the role of despair in America's opioid crisis," Brookings Institution, October 15, 2019, www.brookings.edu/policy2020 /votervital/how-can-policy-address-the-opioid-crisis-and-despair-in-america/.

4. See, for example: Carl Benedikt Frey and Michael A. Osborne, "The future of employment: How susceptible are jobs to computerisation?," Oxford Martin School Programme on Technology and Employment, Working Paper, September 17, 2013, www.oxfordmartin.ox.ac.uk/downloads/academic /future-of-employment.pdf, p. 38.

5. U.S. Bureau of Labor Statistics, "Unemployment rate (UNRATE)," retrieved from Federal Reserve Bank of St. Louis, July 18, 2020, fred.stlouisfed .org/series/UNRATE; Greg Rosalsky, "Are we even close to full employment?," NPR Planet Money, July 2, 2019, www.npr.org/sections/money /2019/07/02/737790095/are-we-even-close-to-full-employment.

6. Organization for Economic Co-operation and Development, "Activity rate: Aged 25–54: Males for the United States (LRAC25MAUSM156S)," retrieved from Federal Reserve Bank of St. Louis, July 17, 2020, fred.stlouisfed .org/series/LRAC25MAUSM156S.

7. "Trends in Social Security Disability Insurance," Social Security Office of Retirement and Disability Policy, Briefing Paper No. 2019-01, August 2019, www.ssa.gov/policy/docs/briefing-papers/bp2019-01.html.

8. U.S. Bureau of Labor Statistics, "Labor force participation rate (CIVPART)," retrieved from Federal Reserve Bank of St. Louis, July 17, 2020, fred.stlouisfed.org/series/CIVPART.

9. U.S. Bureau of Labor Statistics, "Business sector: Real output per hour of all persons (OPHPBS)," retrieved from Federal Reserve Bank of St. Louis, July 22, 2020, fred.stlouisfed.org/series/OPHPBS; U.S. Bureau of Labor Statistics, "Business sector: Real compensation per hour (PRS84006151)," retrieved from Federal Reserve Bank of St. Louis, July 22, 2020, fred.stlouisfed.org /series/PRS84006151.

10. World Bank, "GINI index for the United States (SIPOVGINIUSA)," retrieved from Federal Reserve Bank of St. Louis, July 20, 2020, fred.stlouisfed .org/series/SIPOVGINIUSA.

11. Martha Ross and Nicole Bateman, "Low-wage work is more pervasive than you think, and there aren't enough 'good jobs' to go around," Brookings Institution, November 21, 2019, www.brookings.edu/blog/the -avenue/2019/11/21/low-wage-work-is-more-pervasive-than-you-think-and -there-arent-enough-good-jobs-to-go-around/.

12. "The U.S. Private Sector Job Quality Index (JQI)," accessed July 15, 2020, www.jobqualityindex.com/.

13. Gwynn Guilford, "The great American labor paradox: Plentiful jobs, most of them bad," *Quartz*, November 21, 2019, qz.com/1752676 /the-job-quality-index-is-the-economic-indicator-weve-been-missing/.

14. Elizabeth Redden, "41% of recent grads work in jobs not requiring a degree," *Inside Higher Ed*, February 18, 2020, www.insidehighered.com /quicktakes/2020/02/18/41-recent-grads-work-jobs-not-requiring-degree.

15. "The Phillips curve may be broken for good," *The Economist*, November 1, 2017, www.economist.com/graphic-detail/2017/11/01/the-phillips -curve-may-be-broken-for-good.

16. Jeff Jeffrey, "U.S. companies are rolling in cash, and they're growing increasingly fearful to spend it," *The Business Journals*, December 12, 2018, www.bizjournals.com/bizjournals/news/2018/12/12/u-s-companies-are -hoarding-cash-and-theyre-growing.html.

17. Martin Ford, *Rise of the Robots: Technology and the Threat of a Jobless Future*, Basic Books, 2015, pp. 206–212.

18. Martin Ford, Interview with James Manyika, in *Architects of Intelligence: The Truth about AI from the People Building It*, Packt Publishing, 2018, pp. 285–286.

19. Nir Jaimovich and Henry E. Siu, "Job polarization and jobless recoveries," National Bureau of Economic Research, Working Paper 18334, issued in August 2012, revised in November 2018, www.nber.org/papers/w18334.

20. Jacob Bunge and Jesse Newman, "Tyson turns to robot butchers, spurred by coronavirus outbreaks," *Wall Street Journal*, July 9, 2020, www.wsj.com/articles/meatpackers-covid-safety-automation-robots-coronavirus-11594303535.

21. Miso Robotics, "White Castle selects Miso Robotics for a new era of artificial intelligence in the fast food industry," Press Release Newswire, July 14, 2020, www.prnewswire.com/news-releases/white-castle-selects-miso-robotics-for-a-new-era-of-artificial-intelligence-in-the-fast-food-industry-301092746.html.

22. James Manyika, Susan Lund, Michael Chui, et al., "Jobs lost, jobs gained: What the future of work will mean for jobs, skills, and wages," McKinsey Global Institute, November 28, 2017, www.mckinsey.com/featured-insights/future-of-work/jobs-lost-jobs-gained-what-the-future-of-work-will-mean-for-jobs-skills-and-wages.

23. Ferris Jabr, "Cache cab: Taxi drivers' brains grow to navigate London's streets," *Scientific American*, December 8, 2011, www.scientificamerican.com/article/london-taxi-memory/.

24. Kate Conger, "Facebook starts planning for permanent remote workers," *New York Times*, May 21, 2020, www.nytimes.com/2020/05/21/technology/facebook-remote-work-coronavirus.html.

25. Alexandre Tanzi, "Gloom grips U.S. small businesses, with 52% predicting failure," *Bloomberg*, May 6, 2020, www.bloomberg.com/news/articles/2020-05-06/majority-of-u-s-small-businesses-expect-to-close-survey-says.

26. Alfred Liu, "Robots to cut 200,000 U.S. bank jobs in next decade, study says," *Bloomberg*, October 1, 2019, www.bloomberg.com/news/articles/2019-10-02/robots-to-cut-200-000-u-s-bank-jobs-in-next-decade-study-says.

27. Jack Kelly, "Artificial intelligence is superseding well-paying Wall Street jobs," *Forbes*, December 10, 2019, www.forbes.com/sites/jackkelly/2019/12/10/artificial-intelligence-is-superseding-well-paying-wall-street-jobs/.

28. "Top healthcare chatbots startups," Tracxn, October 20, 2020, tracxn.com/d/trending-themes/Startups-in-Healthcare-Chatbots.

29. Celeste Barnaby, Satish Chandra and Frank Luan, "Aroma: Using machine learning for code recommendation," Facebook AI Blog, April 4, 2019, ai.facebook.com/blog/aroma-ml-for-code-recommendation/.

30. Will Douglas Heaven, "OpenAI's new language generator GPT-3 is shockingly good—and completely mindless," *MIT Technology Review*, July 20, 2020, www.technologyreview.com/2020/07/20/1005454/openai-machine-learning-language-generator-gpt-3-nlp/.

31. Jacques Bughin, Jeongmin Seong, James Manyika, et al., "Notes from the AI frontier: Modeling the impact of AI on the world economy," McKinsey

Global Institute, Discussion Paper, September 2018, www.mckinsey.com/~/media /McKinsey/Featured%20Insights/Artificial%20Intelligence/Notes%20from% 20the%20frontier%20Modeling%20the%20impact%20of%20AI%20on%20 the%20world%20economy/MGI-Notes-from-the-AI-frontier-Modeling-the -impact-of-AI-on-the-world-economy-September-2018.ashx.

32. Anand S. Rao and Gerard Verweij, "Sizing the prize: What's the real value of AI for your business and how can you capitalise?," PwC, October 2018, www.pwc.com/gx/en/issues/analytics/assets/pwc-ai-analysis-sizing-the -prize-report.pdf.

33. Bughin et al., "Notes from the AI frontier: Modeling the impact of AI on the world economy," p 3.

CHAPTER 7. CHINA AND THE RISE OF THE AI SURVEILLANCE STATE

1. Chris Buckley, Paul Mozur and Austin Ramzy, "How China turned a city into a prison," *New York Times*, April 4, 2019, www.nytimes.com/inter active/2019/04/04/world/asia/xinjiang-china-surveillance-prison.html.

2. James Vincent, "Chinese netizens spot AI books on president Xi Jin-ping's bookshelf," *The Verge*, January 3, 2018, www.theverge.com/2018 /1/3/16844364/china-ai-xi-jinping-new-years-speech-books.

3. Tom Simonite, "China is catching up to the US in AI research— fast," *Wired*, March 13, 2019, www.wired.com/story/china-catching-up-us -in-ai-research/.

4. Robust Vision Challenge website, accessed July 25, 2020, www.robust vision.net/rvc2018.php.

5. National University of Defense Technology website, accessed July 25, 2020, english.nudt.edu.cn/About/index.htm.

6. Nicolas Thompson and Ian Bremmer, "The AI Cold War that threatens us all," *Wired*, October 23, 2018, www.wired.com/story/ai-cold-war-china -could-doom-us-all/.

7. Alex Hern, "China censored Google's AlphaGo match against world's best Go player," *The Guardian*, May 24, 2017, www.theguardian .com/technology/2017/may/24/china-censored-googles-alphago-match -against-worlds-best-go-player.

8. China's State Council, "New Generation Artificial Intelligence Devel-opment Plan," issued by China's State Council on July 20, 2017, translated by Graham Webster, Rogier Creemers, Paul Triolo and Elsa Kania, New America Foundation, August 1, 2017, www.newamerica.org/cybersecurity -initiative/digichina/blog/full-translation-chinas-new-generation-artificial -intelligence-development-plan-2017/. (Original Chinese government docu-ment: www.gov.cn/zhengce/content/2017-07/20/content_5211996.htm.)

9. Lai Lin Thomala, "Number of internet users in China 2008–2020," Statista, April 30, 2020, www.statista.com/statistics/265140/number-of -internet-users-in-china/.

10. Lai Lin Thomala, "Penetration rate of internet users in China 2008–2020," Statista, April 30, 2020, www.statista.com/statistics/236963 /penetration-rate-of-internet-users-in-china/.

11. Rachel Metz, "Baidu could beat Google in self-driving cars with a totally Google move," *MIT Technology Review*, January 8, 2018, www .technologyreview.com/2018/01/08/146351/baidu-could-beat-google -in-self-driving-cars-with-a-totally-google-move/.

12. Jon Russell, "Former Microsoft executive and noted AI expert Qi Lu joins Baidu as COO," TechCrunch, January 17, 2017, techcrunch.com /2017/01/16/qi-lu-joins-baidu-as-coo/.

13. Martin Ford, Interview with Demis Hassabis, in *Architects of Intelligence: The Truth about AI from the People Building It*, Packt Publishing, 2018, p. 179.

14. Field Cady and Oren Etzioni, "China may overtake US in AI research," Allen Institute for AI Blog, March 13, 2019, medium.com/ai2-blog /china-to-overtake-us-in-ai-research-8b6b1fe30595.

15. Jeffrey Ding, "Deciphering China's AI dream: The context, components, capabilities, and consequences of China's strategy to lead the world in AI," Future of Humanity Institute, University of Oxford, March 2018, www .fhi.ox.ac.uk/wp-content/uploads/Deciphering_Chinas_AI-Dream.pdf.

16. Jeffrey Ding, "China's current capabilities, policies, and industrial ecosystem in AI: Testimony before the U.S.-China Economic and Security Review Commission Hearing on Technology, Trade, and Military-Civil Fusion: China's Pursuit of Artificial Intelligence, New Materials, and New Energy," June 7, 2019, www.uscc.gov/sites/default/files/June%207%20Hearing_Panel%201 _Jeffrey%20Ding_China%27s%20Current%20Capabilities%2C%20 Policies%2C%20and%20Industrial%20Ecosystem%20in%20AI.pdf.

17. Kai-Fu Lee, "What China can teach the U.S. about artificial intelligence," *New York Times*, September 22, 2018, www.nytimes.com/2018/09/22 /opinion/sunday/ai-china-united-states.html.

18. Kathrin Hille and Richard Waters, "Washington unnerved by China's 'military-civil fusion,'" *Financial Times*, November 7, 2018, www.ft.com /content/8dcb534c-dbaf-11e8-9f04-38d397e6661c.

19. Scott Shane and Daisuke Wakabayashi, "'The Business of War': Google employees protest work for the Pentagon," *New York Times*, April 4, 2018, www.nytimes.com/2018/04/04/technology/google-letter-ceo-pentagon -project.html.

20. Tom Simonite, "Behind the rise of China's facial-recognition giants," *Wired*, September 3, 2019, www.wired.com/story/behind-rise-chinas -facial-recognition-giants/.

21. Paul Mozur and Aaron Krolik, "A surveillance net blankets China's cities, giving police vast powers," *New York Times*, December 17, 2019, www .nytimes.com/2019/12/17/technology/china-surveillance.html.

22. Amy B. Wang, "A suspect tried to blend in with 60,000 concertgoers. China's facial-recognition cameras caught him," *Washington Post*, April 13, 2018, www.washingtonpost.com/news/worldviews/wp/2018/04/13/china -crime-facial-recognition-cameras-catch-suspect-at-concert-with-60000 -people/.

23. Paul Mozur, "Inside China's dystopian dreams: A.I., shame and lots of cameras," *New York Times*, July 8, 2018, nytimes.com/2018/07/08/business /china-surveillance-technology.html.

24. Paul Moser, "One month, 500,000 face scans: How China is using A.I. to profile a minority," *New York Times*, April 14, 2019, www. nytimes.com/2019/04/14/technology/china-surveillance-artificial-intelli gence-racial-profiling.html.

25. Ibid.

26. Simina Mistreanu, "Life inside China's social credit laboratory," *Foreign Policy*, April 3, 2018, foreignpolicy.com/2018/04/03/life-inside-chinas -social-credit-laboratory/.

27. Echo Huang, "Garbage-sorting violators in China now risk being punished with a junk credit rating," *Quartz*, January 8, 2018, qz.com/1173975 /garbage-sorting-violators-in-china-risk-getting-a-junk-credit-rating/.

28. Maya Wang, "China's chilling 'social credit' blacklist," Human Rights Watch, December 12, 2017, www.hrw.org/news/2017/12/13/chinas -chilling-social-credit-blacklist.

29. Nicole Kobie, "The complicated truth about China's social credit system," *Wired*, June 7, 2019, www.wired.co.uk/article/china-social-credit -system-explained.

30. Steven Feldstein, "The global expansion of AI surveillance," Carnegie Endowment for International Peace, September 17, 2019, carnegie endowment.org/2019/09/17/global-expansion-of-ai-surveillance-pub-79847.

31. Yuan Yang and Madhumita Murgia, "Facial recognition: How China cornered the surveillance market," *Financial Times*, December 6, 2019, www .ft.com/content/6f1a8f48-1813-11ea-9ee4-11f260415385.

32. Russell Brandon, "The case against Huawei, explained," *The Verge*, May 22, 2019, www.theverge.com/2019/5/22/18634401/huawei-ban-trump -case-infrastructure-fears-google-microsoft-arm-security.

33. Will Knight, "Trump's latest salvo against China targets AI firms," *Wired*, October 9, 2019, www.wired.com/story/trumps-salvo-against-china -targets-ai-firms/.

34. Kashmir Hill, "The secretive company that might end privacy as we know it," *New York Times*, January 18, 2020, www.nytimes.com/2020/01/18 /technology/clearview-privacy-facial-recognition.html.

35. Ibid.

36. Ibid.

37. Ryan Mac, Caroline Haskins and Logan McDonald, "Clearview's facial recognition app has been used by the Justice Department, ICE, Macy's, Walmart, and the NBA," *BuzzFeed News*, February 27, 2020, www.buzzfeed news.com/article/ryanmac/clearview-ai-fbi-ice-global-law-enforcement.

38. Alfred Ng and Steven Musil, "Clearview AI hit with cease-and-desist from Google, Facebook over facial recognition collection," CNET, February 5, 2020, www.cnet.com/news/clearview-ai-hit-with-cease-and-desist-from-google -over-facial-recognition-collection/.

39. Zack Whittaker, "Apple has blocked Clearview AI's iPhone app for violating its rules," *TechCrunch*, February 28, 2020, techcrunch.com/2020/02/28 /apple-ban-clearview-iphone/.

40. Nick Statt, "ACLU sues facial recognition firm Clearview AI, calling it a 'nightmare scenario' for privacy," *The Verge*, May 28, 2020, www.the verge.com/2020/5/28/21273388/aclu-clearview-ai-lawsuit-facial-recognition -database-illinois-biometric-laws.

41. Paul Bischoff, "Surveillance camera statistics: Which cities have the most CCTV cameras?," Comparitech, August 1, 2019, www.comparitech .com/vpn-privacy/the-worlds-most-surveilled-cities/.

42. "Met Police to deploy facial recognition cameras," BBC, January 30, 2020, www.bbc.com/news/uk-51237665.

43. Clare Garvie, Alvaro Bedoya and Jonathan Frankle, "The perpetual line-up: Unregulated police face recognition in America," Georgetown Law Center on Privacy and Technology, October 18, 2016, www.perpetuallineup .org/.

44. "Met Police to deploy facial recognition cameras."

45. London Real, "Jonathan Haidt—Free range kids: How to give your children more freedom (video)," October 27, 2018, www.youtube.com /watch?v=GPTei2sroIk.

46. Isabella Garcia, "Can facial recognition overcome its racial bias?," *Yes! Magazine*, April 16, 2020, www.yesmagazine.org/social-justice/2020/04/16 privacy-facial-recognition/.

47. Sasha Ingber, "Facial recognition software wrongly identifies 28 lawmakers as crime suspects," NPR, July 26, 2018, www.npr.org/2018/07/26 /632724239/facial-recognition-software-wrongly-identifies-28-lawmakers -as-crime-suspects.

48. Patrick Grother, Mei Ngan and Kayee Hanaoka, "Face Recognition Vendor Test (FRVT) Part 3: Demographic effects," National Institute of Standards and Technology, December 2019, nvlpubs.nist.gov/nistpubs/ir/2019 /NIST.IR.8280.pdf.

49. Garcia, "Can facial recognition overcome its racial bias?"

50. Amy Hawkins, "Beijing's big brother tech needs African faces," *Foreign Policy*, July 24, 2018, foreignpolicy.com/2018/07/24/beijings-big-brother-tech -needs-african-faces/.

CHAPTER 8. AI RISKS

1. "Fake voices 'help cyber-crooks steal cash,'" BBC News, July 8, 2019, www.bbc.com/news/technology-48908736.

2. Martin Giles, "The GANfather: The man who's given machines the gift of imagination," *MIT Technology Review*, February 21, 2018, www.tech-nologyreview.com/2018/02/21/145289/the-ganfather-the-man-whos-given -machines-the-gift-of-imagination/.

3. James Vincent, "Watch Jordan Peele use AI to make Barack Obama deliver a PSA about fake news," *The Verge*, April 17, 2018, www.theverge .com/tldr/2018/4/17/17247334/ai-fake-news-video-barack-obama-jordan peele-buzzfeed.

4. Sensity, "The state of deepfakes 2019: Landscape, threats, and impact," September 2019, sensity.ai/reports/.

5. Ian Sample, "What are deepfakes—and how can you spot them?," *The Guardian*, January 13, 2020, www.theguardian.com/technology/2020/jan/13 /what-are-deepfakes-and-how-can-you-spot-them.

6. Lex Fridman, "Ian Goodfellow: Generative Adversarial Networks (GANs)," Artificial Intelligence Podcast, episode 19, April 18, 2019, lexfridman .com/ian-goodfellow/. (Video and audio podcast available.)

7. J.J. McCorvey, "This image-authentication startup is combating faux social media accounts, doctored photos, deep fakes, and more," *Fast Company*, February 19, 2019, www.fastcompany.com/90299000/truepic -most-innovative-companies-2019.

8. Ian Goodfellow, Nicolas Papernot, Sandy Huang, et al., "Attacking machine learning with adversarial examples," OpenAI Blog, February 24, 2017, openai.com/blog/adversarial-example-research/.

9. Anant Jain, "Breaking neural networks with adversarial attacks," To-wards Data Science, February 9, 2019, towardsdatascience.com/breaking -neural-networks-with-adversarial-attacks-f4290a9a45aa.

10. Ibid.

11. *Slaughterbots*, released November 12, 2017, Space Digital, www .youtube.com/watch?reload=9&v=9CO6M2HsoIA.

12. Stuart Russell, "Building a lethal autonomous weapon is easier than building a self-driving car. A new treaty is necessary," *The Security Times*, February 2018, www.the-security-times.com/building-a-lethal-autonomous -weapon-is-easier-than-building-a-self-driving-car-a-new-treaty-is-necessary/.

13. Martin Ford, Interview with Stuart Russell, in *Architects of Intelligence: The Truth about AI from the People Building It*, Packt Publishing, 2018, p. 59.

14. "Country views on killer robots," Campaign to Stop Killer Robots, August 21, 2019, www.stopkillerrobots.org/wp-content/uploads/2019/08 /KRC_CountryViews21Aug2019.pdf.

15. "Russia, United States attempt to legitimize killer robots," Campaign to Stop Killer Robots, August 22, 2019, www.stopkillerrobots.org/2019/08 /russia-united-states-attempt-to-legitimize-killer-robots/.

16. Zachary Kallenborn, "Swarms of mass destruction: The case for declaring armed and fully autonomous drone swarms as WMD," Modern War Institute, May 28, 2020, mwi.usma.edu/swarms-mass-destruction-case-declaring -armed-fully-autonomous-drone-swarms-wmd/.

17. Kris Osborn, "Here come the Army's new class of 10-ton robots," *National Interest*, May 21, 2020, nationalinterest.org/blog/buzz/here-come -armys-new-class-10-ton-robots-156351.

18. Rachel England, "The US Air Force is preparing a human versus AI dogfight," *Engadget*, June 8, 2020, www.engadget.com/the-air-force-will-pit-an -autonomous-fighter-drone-against-a-pilot-121526011.html.

19. Kris Osborn, "Robot vs. robot war? Now China has semi-autonomous fighting ground robots," *National Interest*, June 15, 2020, nationalinterest .org/blog/buzz/robot-vs-robot-war-now-china-has-semi-autonomous-fighting -ground-robots-162782.

20. Neil Johnson, Guannan Zhao, Eric Hunsader, et al., "Abrupt rise of new machine ecology beyond human response time," *Nature Scientific Reports*, volume 3, article number 2627 (2013), September 11, 2013, www.nature.com /articles/srep02627.

21. Ford, Interview with Stuart Russell, in *Architects of Intelligence*, p. 59.

22. Jeffrey Dastin, "Amazon scraps secret AI recruiting tool that showed bias against women," Reuters, October 10, 2018, www.reuters.com/article /us-amazon-com-jobs-automation-insight/amazon-scraps-secret-ai-recruiting -tool-that-showed-bias-against-women-idUSKCN1MK08G.

23. Julia Angwin, Jeff Larson, Surya Mattu and Lauren Kirchner, "Machine bias," *Propublica*, May 23, 2016, www.propublica.org/article/machine -bias-risk-assessments-in-criminal-sentencing.

24. Ibid.

25. Ford, Interview with James Manyika, in *Architects of Intelligence*, p. 279.

26. Ford, Interview with Fei-Fei Li, in *Architects of Intelligence*, p. 157.

27. Stephen Hawking, Stuart Russell, Max Tegmark and Frank Wilczek, "Stephen Hawking: 'Transcendence looks at the implications of artificial intelligence—but are we taking AI seriously enough?,'" *The Independent*, May 1, 2014, www.independent.co.uk/news/science/stephen-hawking-transcendence

-looks-at-the-implications-of-artificial-intelligence-but-are-we-taking-ai
-seriously-enough-9313474.html.

28. Nick Bostrom, *Superintelligence: Paths, Dangers, Strategies*, Oxford University Press, 2014, p. vii.

29. Matt McFarland, "Elon Musk: 'With artificial intelligence we are summoning the demon,'" *Washington Post*, October 24, 2014, www.washingtonpost .com/news/innovations/wp/2014/10/24/elon-musk-with-artificial-intelligence -we-are-summoning-the-demon/.

30. Sam Harris, "Can we build AI without losing control over it? (video)," TED Talk, June 2016, www.ted.com/talks/sam_harris_can_we_build_ai_with out_losing_control_over_it?language=en.

31. Irving John Good, "Speculations concerning the first ultraintelligent machine," *Advanced in Computers*, volume 6, pp. 31–88 (1965), vtechworks .lib.vt.edu/bitstream/handle/10919/89424/TechReport05-3.pdf.

32. Jesselyn Cook, "Hundreds of people share stories about falling down YouTube's recommendation rabbit hole," *Huffington Post*, October 15, 2019, www.huffpost.com/entry/youtube-recommendation-rabbit-hole-mozilla_n _5da5c470e4b08f3654912991.

33. Stuart Russell, *Human Compatible: Artificial Intelligence and the Problem of Control*, Viking, 2019, pp. 173–177.

34. Stuart Russell, "How to stop superhuman A.I. before it stops us," *New York Times*, October 8, 2019, www.nytimes.com/2019/10/08/opinion/artificial -intelligence.html.

35. Ford, Interview with Rodney Brooks, in *Architects of Intelligence*, pp. 440–441.

CONCLUSION: TWO AI FUTURES

1. Rebecca Heilweil, "Big tech companies back away from selling facial recognition to police," *Recode*, June 11, 2020, www.vox.com/recode/2020 /6/10/21287194/amazon-microsoft-ibm-facial-recognition-moratorium-police.

2. Joseph Zeballos-Roig, "Kamala Harris supports $2,000 monthly stimulus checks to help Americans claw out of pandemic ruin—and she's long backed plans for Democrats to give people more money," *Business Insider*, August 15, 2020, www.businessinsider.com/kamala-harris-biden-monthly -stimulus-checks-economic-policy-support-vice-2020-8.

3. Bob Berwyn, "What does '12 years to act on climate change' (now 11 years) really mean?," *Inside Climate News*, August 27, 2019, insideclimatenews.org /news/27082019/12-years-climate-change-explained-ipcc-science-solutions.

4. Bill Gates, "COVID-19 is awful. Climate change could be worse," Gates Notes, August 4, 2020, www.gatesnotes.com/Energy/Climate-and-COVID-19.

5. Bill Gates, "Climate change and the 75% problem," Gates Notes, October 17, 2018, www.gatesnotes.com/Energy/My-plan-for-fighting-climate-change.

6. Nicholas Bloom, Charles I. Jones, John Van Reenen and Michael Webb, "Are ideas getting harder to find?," *American Economic Review*, volume 110, issue 4, pp. 1104–1144 (April 2020), www.aeaweb.org/articles?id=10.1257/aer.20180338, p. 1138.

7. Mark Aguiar, Mark Bils, Kerwin Kofi Charles and Erik Hurst, "Leisure luxuries and the labor supply of young men," National Bureau of Economic Research, Working Paper 23552, June 2017, www.nber.org/papers/w23552.

INDEX

MARTIN FORD is a futurist and the author of the *New York Times* bestseller *Rise of the Robots*, which won the *Financial Times* Business Book of the Year Award; *Architects of Intelligence*; and *The Lights in the Tunnel*. He is also the founder of a Silicon Valley–based software development firm. His TED Talk on the impact of artificial intelligence on society has been viewed over 3 million times, and his writing has appeared in the *New York Times*, *Fortune*, *Forbes*, the *Atlantic*, the *Washington Post*, *Harvard Business Review*, the *Guardian*, and the *Financial Times*. Ford is a sought-after speaker and a leading expert on artificial intelligence. He lives in Sunnyvale, California.

Twitter: @MFordFuture